Atlas of Early Zebrafish Brain Development

A Tool for Molecular Neurogenetics

Atlas of Early Zebrafish Brain Development

A Tool for Molecular Neurogenetics

Thomas Mueller
and
Mario F. Wullimann

CNRS
Institute of Neurobiology A. Fessard
Development, Evolution, Plasticity of the Nervous System
Research Unit 2197, Avenue de la Terrasse
F-91198 Gif-sur-Yvette, France
(Thomas.Mueller@iaf.cnrs-gif.fr *or*
Mario.Wullimann@iaf.cnrs-gif.fr *or*
Wullimann@uni-bremen.de)

ELSEVIER
Amsterdam – Boston – Dusseldorf – London – New York – Oxford
Paris – San Diego – San Francisco – Singapore – Sydney – Tokyo
2005

ELSEVIER B.V.
Radarweg 29
P.O. Box 211, 1000 AE Amsterdam
The Netherlands

ELSEVIER Inc.
525 B Street, Suite 1900
San Diego, CA 92101-4495
USA

ELSEVIER Ltd
The Boulevard, Langford Lane
Kidlington, Oxford OX5 1GB
UK

ELSEVIER Ltd
84 Theobalds Road
London WC1X 8RR
UK

First edition 2005

British Library Cataloguing in Publication Data
Mueller, Thomas.
Atlas of early zebrafish brain development : a tool for molecular neurogenetics
1. Zebra danio-Development -Atlases 2. Brain - Anatomy - Atlases 3. Neurogenetics - Atlases
I.Title II.Wullimann, Mario F.
573.8'617482

ISBN 0444517383

Library of Congress Cataloguing-in- Publication Data
Mueller, Thomas.
Atlas of early zebrafish brain development : a tool for molecular neurogenetics / Thomas Mueller and Mario F. Wullimann.-1st ed.
p. cm.
Includes bibliographical references and index.
ISBN 0-444-51738-3 (alk. paper)
1. Zebra danio -Anatomy-Atlases. 2. Brain-Anatomy-Atlases. 3. Neurogenetics. I. Wullimann, Mario F. II. Title.
QL638.C94M84 2005
573.8'617482-dc22

2004066172

♾ The paper used in this publication meets the requirements of ANSI/NISO Z39.48-1992 (Permanence of Paper).
Printed in The Netherlands.

Preface

No one doubts that methodological revolutions shape the direction of science. For instance, the invention of the microscope led almost directly to the recognition of the cell as the basic module of all organisms. Equally decisive for the scientific progress are ideas and concepts which sometimes predate their confirmation by adequate methodology. These two closely intertwined processes shape past and present scientific endeavors and are part of human culture created by the efforts of people interconnected by a common interest in similar questions over many decades. The admiration for methodological innovations and the respect for scientific ideas or concepts both represent important foundations of science.

Therefore, Chapter 1 of this book puts into perspective historically important contributions to the field of early vertebrate central nervous system development. With this, we wish to clarify how the cellular–molecular data delivered in Chapter 2 relate to the larger neurobiological context of neurogenesis in early brain development. The second chapter delivers the actual atlas pages that the readership may use pragmatically to identify certain early zebrafish brain areas. Chapter 3 takes up the dynamical aspects of how the atlas may be used, be it to elucidate locally differing dynamics of proliferation, determination and differentiation of neural cells or to implement additional gene expression data. Finally, Chapter 4 delivers a comparative perspective of gene expression patterns between major model animals, especially between mouse and zebrafish.

In addition to delivering a refined early neuroanatomy of the zebrafish brain, the cellular and molecular genetic nature of the data presented in this book allows a starting point for understanding the local differences in the distribution of proliferative, early neuronally determined and differentiating cells and their underlying neurogenetic pathways. The book also is intended to facilitate the use of the zebrafish model animal in many future neurobiologically oriented investigations in basic as well as biomedically relevant research projects.

Thomas Mueller
Mario F. Wullimann
Gif-sur-Yvette, France, December 2004

Acknowledgements

We would like to thank the following colleagues for help during the generation of this book: Philippe Vernier (Director IAF; Gif-sur-Yvette) for his most generous support of this project, many insightful discussions and for critically reading the manuscript, Patrick Blader (Toulouse, France) and Uwe Strähle (Karlsruhe, Germany), Michael Lardelli (Adelaide, Australia), and Eric Weinberg (Philadelphia, USA) for various plasmids, Elke Rink (Bremen, Germany) for providing the Pax6 immunostain and innumerable other support, Rachel McDonald, Ingvild Mikkola, and Steve Wilson (London, UK) for the Pax6 antibody; Franck Bourrat, Catherine Pasqualini, Jean-Stéphane Joly, Robert Luedtke, Heather McLean and Sylvie Rétaux, as well as Alessandro Alunni, Sophie Callier, Eva Maria Candal, Jacob Engelmann, Frédéric Moret, Joana Osorio, Marina Snapyan, Finn-Arne Weltzien, and Maryline Blin and Laurent Legendre for friendly support in Gif-sur-Yvette. Special thanks are also due to Patrick Laurenti, Didier Casane, and Veronique Borday-Birraux for kindly providing access to their zebrafish facility (Gif-sur-Yvette), to Odile Lecquyer for gently helping us to navigate through the French administration system and Jean-Paul Bouillot for the original figure printing. Last but not least, we are deeply grateful to our families and friends who believed in our project even in difficult times. Thomas Mueller especially thanks his mother Helga Mueller, his grandmother Helene de Laar, as well as Nicole Renken; Mario Wullimann his wife Kristina Wullimann and children Lea and Jan, his parents Sonja and Franz Wullimann, and his parents-in-law Klara and Alois Sebralla.

Contents

Vertebrate Central Nervous System Development

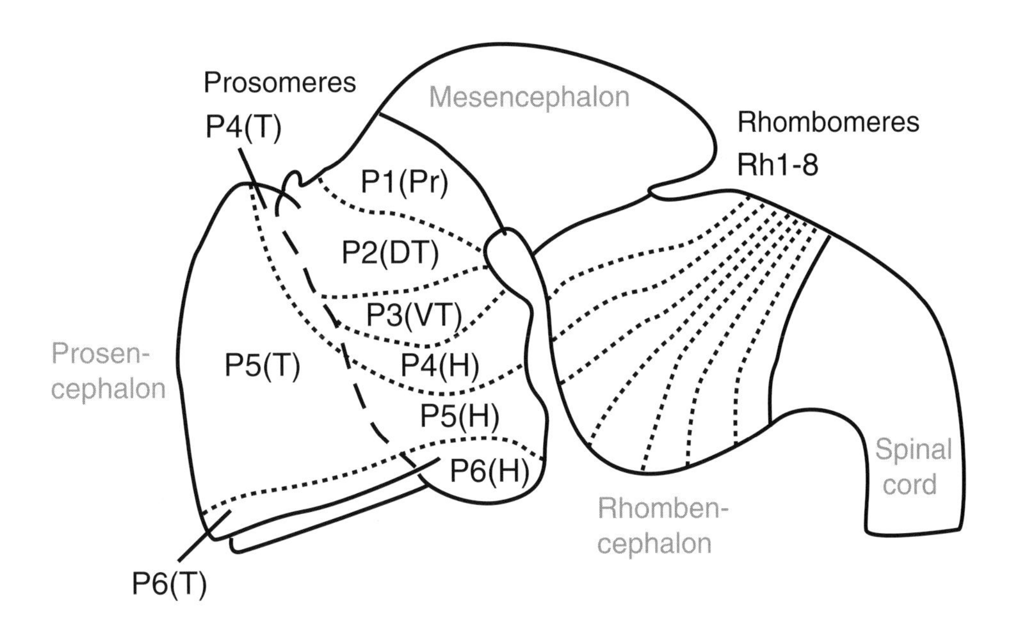

Chapter 1

Introduction

1. Vertebrate Central Nervous System Development

1.1. Introductory Remarks: The Order of the Universe

The individual universe each human being carries around mentally is a product of our brain, which has been said to qualify as the most complex existing object in the world. How does the brain come into existence during development? The ontogeny of the central nervous system (CNS: brain and spinal cord) is itself a highly complex morphogenetic process characterized by a strong aspect of spatiotemporal order observed similarly in all vertebrates during development. The early vertebrate embryo exhibits three embryonic cellular sheets called germ layers, the most peripheral one being the ectoderm. Its dorsal portion represents the neuroectoderm (the future nervous system), which segregates early from the ventrolaterally lying general ectoderm (the future epidermal skin and its derivatives). Subsequently, the central main portion of the neuroectoderm, the neural plate, is "swallowed" by the general ectoderm through the process of neurulation. As a consequence, the neural plate separates from the general ectoderm and descends into the deep of the embryo, where it develops into a hollow neural tube surrounding a cerebrospinal fluid- (liquor-) filled ventricle. Meanwhile, the general ectoderm closes the temporarily opened dorsal backside of the embryo. The apparent formation of three initial (forebrain, midbrain, hindbrain) and subsequent five brain vesicles (telencephalon, diencephalon, mesencephalon, metencephalon, myelencephalon; but see discussion in the next section) along the anteroposterior axis somewhat hides the parallel establishment of finer subdivisions called neuromeres (rhombomeres in the hindbrain, prosomeres in the forebrain). Next is the differentiation of functional subdivisions of the brain and spinal cord (for example, the optic tectum/superior colliculus, a midbrain sensorimotor integration center, or the hypothalamus, the major control center of visceral processes).

These events of vertebrate CNS formation are accompanied by the emigration of the most lateral portions of the neuroectoderm, the neural folds (*Neuralwülste*) and some directly adjacent neuroectodermal regions, which are not incorporated into the neural tube during neurulation. These neuroectodermal components, that is, the neural crest (in head and body trunk) and the placodes (only in head), eventually migrate to their final destination in the adult body where they form (among other tissues) practically all sensory organs and their innervating neurons and nerves (somatic and visceral peripheral nervous system; PNS).

For many decades, these complex developmental events leading to the formation of a nervous system have been investigated in various vertebrates, especially in particular model animals, such as the African clawed frog *Xenopus*, the chick, the mouse and rat, and, most recently, the zebrafish, on every possible level, from molecular aspects over cellular features to histological/morphological phenomena. Thus, a wealth of scientific literature is available, ready to satisfy almost any hunger for information.

What then is to follow in the present book? It integrates knowledge on CNS development coming from classical developmental studies with recent molecular data related to neuro- and gliogenesis. We will demonstrate the detailed spatiotemporal order of early postembryonic zebrafish brain development in its entirety, using molecular and cellular markers of neuro- and gliogenesis to visualize local differences, embedded in a holistic morphogenetic context. This combination of old and new knowledge results in a neuroanatomically based molecular atlas of neural development in the zebrafish brain. Hopefully, such information—beyond delivering a refined molecular neuroanatomy—will prove to be enlightening in the future elucidation of neurogenetic pathways and their mechanisms. Thus, the promise that the zebrafish model holds regarding its more effective future use in biomedical related neurobiological research might be brought a step forward with this book.

1.2. Major Developmental Stages of the Vertebrate Neural Tube

What are the real morphological or functional units of the vertebrate brain and how do they develop? Before going into a more detailed description of the development of finer brain subdivisions (see section 1.3), we will shortly examine the early events of brain morphogenesis relating to the apparent emergence of vesicles and neuromeres, as well as of longitudinal zones. We shall highlight some historically critical cognitive steps during this scientific process bearing directly on the understanding of neuro- and gliogenesis as outlined with molecular markers later in this book. An excellent in-depth historical account on the understanding of vertebrate brain morphogenesis has recently been given by Nieuwenhuys (1998a,b,c).

1.2.1. Vesicles, Neuromeres and Longitudinal Zones Now and Then

The classical description of a sequential appearance of two, three and then five brain vesicles (i.e., transverse elements) along the anteroposterior vertebrate neural tube axis dates back to von Baer (1828), but became more popular in the early 20th century (von Kupffer 1906; Johnston 1909). The two vesicle stage comprises a combined forebrain (prosencephalon) and midbrain (mesencephalon) vesicle which is set apart by a vertical neural tube constriction from the hindbrain (rhombencephalon) vesicle. This is followed by the three vesicle stage displaying forebrain, midbrain and hindbrain vesicles (see fig. 1) and, finally, by the five vesicle stage, consisting of telencephalon, diencephalon (together forming the forebrain), mesencephalon, metencephalon and myelencephalon (the latter two forming the rhombencephalon).

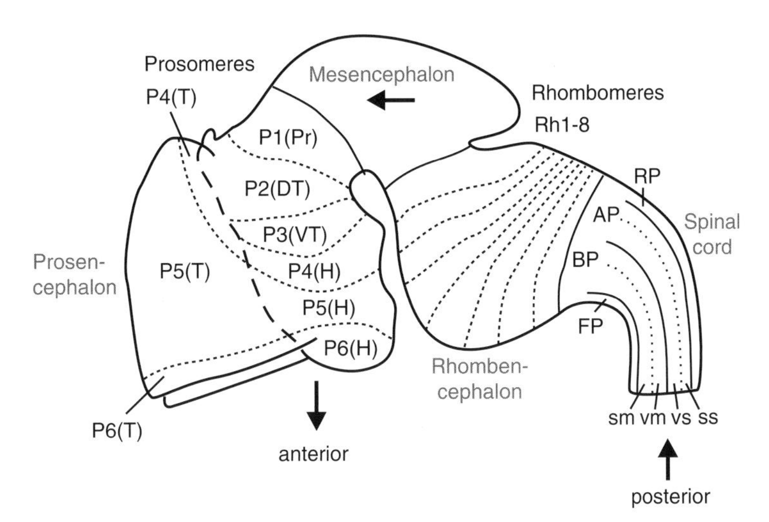

Fig. 1. Schematic lateral view of early mouse brain (E 12.5–13.5) shows prosomeric interpretation (Puelles and Rubenstein 1993), including transverse (neuromeres) and longitudinal (columns) elements. Arrows designate anteroposterior axis of neural tube. Abbreviations: AP, alar plate; BP, basal plate; DT, dorsal thalamus (thalamus); FP, floor plate; H, hypothalamus; P1–P6, prosomeres 1–6 (for more details see text); Pr, pretectum; Rh1–Rh8, rhombomeres 1–8; RP, roof plate; sm, somatomotor column; ss, somaotosensory column; T, telencephalon; vm, visceromotor column; vs, viscerosensory column; VT, ventral thalamus (prethalamus).

Further subdivision of both forebrain and hindbrain into two separable (transverse) vesicles each along the anteroposterior axis may be regarded as an epiphenomenon, visible only in mammalian embryos. These show an enormous early growth of the telencephalic hemispheres. This hides the fact that the most rostral neural tube in reality includes dorsally the telencephalon and ventrally the hypothalamus, rendering telencephalic and diencephalic vesicles obsolete. Also, the somewhat later, characteristically mammalian emergence of a large cerebellum and ventral pontine region led to the impression that the medulla oblongata develops from two separable vesicles, the metencephalic one exhibiting cerebellum and pons, and a myelencephalic one posteriorly. In reality, the medulla oblongata does not show a boundary between these two postulated vesicles, but is rather divided into a higher number of early segmental entities, the rhombomeres (see below and fig. 1). However, the most anterior rhombomere is indeed characterized by the specialization of a dorsal cerebellum and a ventral pontine region that relays cortical information to the cerebellum, resulting in the misleading impression of a metencephalic vesicle that is separable from a posterior myelencephalic one.

In contrast, the three vesicle stage reflects on fundamental anteroposterior vertebrate brain divisions. Especially, the midbrain–hindbrain boundary has been strongly corroborated in modern developmental neurobiology as a singularly definable boundary and signaling center in the vertebrate brain (Marín and Puelles 1994; Bally-Cuif and Wassef 1995; Brand et al. 1996, Lumsden and Krumlauf 1996; Reifers et al. 1998; Wurst and Bally-Cuif 2001). Similarly, there is clear evidence for cellular lineage restriction (Larsen et al. 2001) and signaling center function (Scholpp and Brand 2003) at the forebrain–midbrain boundary.

Equally important for the understanding of the vertebrate CNS *bauplan*, two important flexures were also historically noted early (His 1888, 1893a), the cephalic one between mesencephalon and prosencephalon and the cervical one between rhombencephalon and spinal cord. This results in an early bending of the anteroposterior axis roughly into an upside-down "U" with the mesencephalon forming the top of this turned around "U" (compare with fig. 1). As we shall see later, the recognition of the adequate course of the anteroposterior axis of the neural tube has important consequences for the correct interpretation of the topology of brain structures.

Whereas the vesicle story just outlined found its way easily into general textbook knowledge, the historically equally early description of neural tube segments or so-called neuromeres did not. Neuromeres represent another finer set of transitory transverse or segmental divisions of the vertebrate neural tube, and they remain, at least partially, controversial to the present day. Vertebrate hindbrain neuromeres (i.e., rhombomeres), set apart from each other by transverse, vertical constrictions, were already described in the 19th century (von Baer 1828; Orr 1887). Later, similar morphological observations led to the interpretation of neuromeres to extend into midbrain (mesomeres) and forebrain as well (prosomeres; see fig. 1; Rendahl 1924; Bergquist 1932; Vaage 1969). However, in these early 20th century studies, the basis for recognizing transverse neuromeres was descriptive (and, even worse, transitory) morphology. Thus, serious opposition regarding the acceptance of neuromeres as the important building blocks of the CNS arose with its growing functional neuroanatomical understanding, especially by the so-called American school which, based on the work of Gaskell (1889), was prominently represented by C.J. Herrick. In this school's well corroborated view, longitudinal functional zones—embryonically represented dorsoventrally in the neural tube as roof, alar, basal and floor plates (His 1888, 1893a,b)—were considered the primary subdivisions of the CNS (see fig. 1). Both the most dorsal (roof plate) and most ventral (floor plate) longitudinal zones have critical developmental roles, for example in mediating dorsoventral polarity of the neural tube or in guiding neuronal differentiation processes. In the adult vertebrate CNS, the roof plate gives rise to dorsal midline structures, such as the epiphysis or the chorioid plexus which covers the rhomboid opening of the medulla oblongata, whereas the floor plate develops for example into ventral midline glial cells.

More critical for Herrick's view of functional vertebrate brain organization are the roles of the two intermediate longitudinal zones, that is the alar and basal plates in the developing and adult vertebrate CNS. These form the characteristic dorsoventral arrangement of four functional longitudinal subzones, i.e., alar plate derived somato- and viscerosensory columns and basal plate derived viscero- and somatomotor columns as prototypically present in the spinal cord (see fig. 1). This dorsoventral spinal cord organization of four sensory and motor zones could be used to explain consistently various local specializations (hypertrophies and reductions) in more anterior brain regions—at least into the midbrain. Thus, the concept of longitudinal functional zones understandably gained much acceptance, still expressed legitimately in recent textbooks.

A related and hotly debated question since is how far anteriorly the vertebrate brain longitudinal zones extend (see Nieuwenhuys (1998a)). Considering alar and basal plates here (leaving aside floor and roof plates), there were two fractions again. Kingsbury (1922) believed that the basal plate tapers out at the mesencephalic–diencephalic boundary and that the alar plate, thus, forms the more anterior brain parts. Similarly, Herrick (1910) had proposed earlier that the basal plate becomes thinner anteriorly, but that it continues into the hypothalamus. Here, Herrick's original sin comes in: he believed that also the more dorsal diencephalic divisions (i.e., ventral thalamus, dorsal thalamus and epithalamus) all represent longitudinal zones (i.e., functionally elaborated alar plate subdivisions that he even saw to continue into the telencephalon), which turned out to be a false concept with long-lasting consequences (see below). Kuhlenbeck (publications between 1926 and 1973; see Nieuwenhuys (1998a)) was supportive of Herrick's longitudinal zone concept (including the diencephalon), but highlighted its *bauplan* nature, disregarding the functional aspect. In contrast, Bergquist and Källén (1954)—similar to His (1888, 1893a,b)—proposed the continuation of both alar and basal plates into the general region of the optic chiasm/preoptic region. Thus, in line with their neuromeric concept, the forebrain prosomeres contained alar and basal plate derivatives. This has the important consequence that dorsal and ventral thalamus (and, pretectum, for that matter) represent transverse—not longitudinal—elements along the anteroposterior neural tube axis, which is a now widely accepted concept (see below).

Starting with the 1980s, vertebrate brain neuromery has been reconsidered and confirmed. However, a deep schisma exists between today's "segmentalists". Mostly dealing with rhombomeres (Hannemann et al. 1988; Holland and Hogan 1988; Keynes and Stern 1988; Lumsden and Keynes 1989; Murphy et al. 1989; Lumsden 1990; Trevarrow et al. 1990; Wilkinson and Krumlauf 1990; Clarke and Lumsden 1993; Lumsden and Krumlauf 1996), many recent researchers stress that rigorous criteria for real transverse (neuromeric) neural tube elements (i.e., containing a piece of roof, alar, basal and floor plates each) are exclusively met by rhombomeres. Among those criteria are glial intersegmental boundaries, repetitive (metameric) segmental generation of comparable neuronal classes and axonal routing in each neuromere, and, most importantly, neuromere-specific cellular lineage restriction (Fraser et al. 1990). These criteria led to the most stringent definition available of a segment or neuromere as constituting a neural *compartment* (Larsen et al. 2001). Strong complementary supporting evidence from molecular genetic studies showed that various regulatory genes, especially those of the Hox B cluster (Graham et al. 1989; Hunt and Krumlauf 1992) or others, such as *Krox20* (Wilkinson et al. 1989a,b; Oxtoby and Jowett 1993) are hierarchically or specifically expressed respecting rhombomere boundaries, thereby confining positional information to the rhombomeres, eventually leading to the adult differing phenotypical appearance of their respective derivatives along the anteroposterior hindbrain axis.

Also forebrain neuromeres have been rediscovered in vertebrates recently, and various numbers of such prosomeres were proposed based on new data (Puelles et al. 1987; Figdor and Stern 1993; Puelles and Rubenstein 1993, 2003). The most prominent contemporary, so-called *neuromeric model* by Puelles and Rubenstein (1993, 2003) suggests an overall segmentation of the vertebrate brain and it shall be discussed in some more detail below. Generally, it may be stated that prosomeres (compare with fig. 1) have not been corroborated equally well using the criteria applied to rhombomeres just mentioned. However, cellular lineage restriction definitely occurs at the forebrain–midbrain boundary, i.e., between pretectum (synencephalon, P1) and mesencephalon (Larsen et al. 2001), and within the zona limitans intrathalamica (Zeltser et al. 2001), a transitory transitional region between dorsal thalamus (posterior parencephalon, P2) and ventral thalamus (anterior parencephalon, P3, prethalamus), thereby separating and characterizing P2 and P3 indirectly as neuromeres. The diencephalon has alternatively been reported to contain four neuromeres (Figdor and Stern 1993) characterized by lineage restriction, i.e., two in the synencephalon (P1), and one each in the dorsal (P2) and ventral thalami (P3). Additional potential prosomere boundaries, especially those anterior to P3 (i.e., in the secondary prosencephalon) remain largely unexplored in the context of cellular lineage restriction.

Furthermore, as in the rhombencephalon, some early active regulatory genes are expressed in specific prosomeres (e.g., *Prox* in P1, *Gbx2* in P2, *Dlx2* in P3; Larsen et al. 2001) or respect at least partial prosomeric boundaries with their expression domains (Simeone et al. 1992a; Boncinelli et al. 1993, 1995; Bulfone et al. 1993a,b; Puelles and Rubenstein 1993, 2003). Our own studies regarding the distribution of early proliferation zones also strongly indicate the presence of three posterior prosomeres, i.e., P1 through P3 (Wullimann and Puelles 1999; Mueller and Wullimann 2002b), while no prosomeres are recognized based on proliferation patterns in the more anterior forebrain (secondary prosencephalon). Also, neurogenic (Notch/Delta) and proneural (basic helix–loop–helix) gene expressions closely parallel brain proliferation patterns, including the prosomeric pattern (P1/P2/P3) in the posterior forebrain (Mueller and Wullimann 2003; Wullimann and Mueller 2004a,b; see below).

In conclusion, there is good evidence for three posterior prosomeres (P1 through P3) in the vertebrate forebrain, with the situation in the more anterior forebrain (secondary prosencephalon) being more elusive. Clearly, rhombomeres match better with criteria for neuromeres/compartments (see above) when compared to prosomeres. However, one must keep in mind that these criteria are only met transitionally during neuromere development and that the most critical criterion, i.e., cell lineage restriction, is at no time completely met even in rhombomeres; there are reports on more than 5% of cells transgressing rhombomere boundaries (Birgbauer and Fraser 1994). Furthermore, overt metameric organization (i.e., similar sets of anteroposteriorly repeated classes of neurons) as seen in the developing hindbrain does not necessarily require segmental neural tube organization during development. Metameric organization, as partially still present in the adult rhombencephalon, is even more pervasive in the spinal cord, which does not exhibit neuromeres (as defined above) during any period of development (Keynes and Stern 1984). Therefore, metamery is apparently not necessarily strictly developmentally correlated with a neuromeric organization. Thus, the forebrain simply may be characterized by very early individual regulation of each of its transverse entities, resulting in a fast variable differentiation of neuromeres and their derivatives.

1.2.2. The Neuromeric Model

Irrespective of the ongoing dispute about the true nature and final numbers of vertebrate neuromeres, the so-called neuromeric model of Puelles and Rubenstein (1993, 2003) integrates and puts into perspective much of the issues raised above, suggesting a vertebrate brain *bauplan* that has been of great heuristic value in many studies since. The correct topological identification of a particular brain structure, and, therefore, its comparative interpretation, critically depends on the concept of the basic vertebrate CNS *bauplan* one uses. Such a *bauplan* involves more than the recognition of transverse elements, it also requires to adequately designate the course of the anteroposterior axis along the neural tube, and, in consequence, of the longitudinal zones. Following the neuromeric model, all central nervous longitudinal zones (floor, basal, alar and roof plates) are principally accepted to extend into the general region of the optic chiasma/ preoptic region. This is now supported by studies on various longitudinally expressed genes (such as for example *sonic hedgehog*) which continue anteriorly to be expressed into the forebrain (Ericson et al. 1995; Shimamura et al. 1995, 1997; Hauptmann and Gerster 2000). Such data clearly corroborate the old observation noted above that the anteroposterior axis of the rostral vertebrate neural tube is considerably deflected ventrally (compare with fig. 1). Understanding the course of longitudinal elements is critical for the interpretation of transverse neural tube units (neuromeres), as the latter must lie perpendicular to the anteroposterior axis. In the neuromeric model of Puelles and Rubenstein (1993), the entire brain is principally divided into transverse zones (each containing a segment of all longitudinal zones), and neuromeres therefore exist not only in the rhombencephalon (rhombomeres) but also in the mesencephalon (mesomeres) and prosencephalon (prosomeres), very much following the ideas of Bergquist and Källén described above. In a recent revision of their model, Puelles and Rubenstein (2003) have modified their initial claim of six prosomeres, suggesting only three posterior diencephalic prosomeres (P1/P2/P3), with the secondary prosencephalon possibly guided by different processes of regionalization, much like the conclusion reached in other recent studies discussed above (e.g., Wullimann and Puelles 1999). Otherwise, we interpret the data presented here within the original neuromeric model (Puelles and Rubenstein 1993), as some of the revisions (Puelles and Rubenstein 2003) are based on information unavailable yet in the zebrafish (see also "A word regarding terminology", Chapter 2).

Despite the insecurity about the exact number of prosomeres, it is important to accept the longitudinal axis proposed in the neuromeric model in order to interpret adequately the topological transformation of longitudinal and transverse elements that occurs during the development of a theoretically straight vertebrate neural tube. A good example is the hypothalamus and the posteriorly adjacent basal diencephalic regions (i.e., retromammillary area, posterior tuberculum, basal synencephalon) which are recognized to represent basal (and floor) plate portions of the forebrain neural tube only when this deflected axis is respected (compare with fig. 1). The corresponding forebrain alar (and roof) plate complements are represented by telencephalon/preoptic region anteriorly and by the dorsal aspects of ventral and dorsal thalami (prethalamus and thalamus of Puelles and Rubenstein 2003) and of pretectum posteriorly. A still widely accepted textbook opinion going back to C.J. Herrick (see above) holds that the diencephalon (i.e., hypothalamus, posterior tuberculum, thalamus, epithalamus, pretectum) is a transverse piece of the neural tube in which the hypothalamus represents the most ventral (i.e., basal) part, and the ventral thalamus, dorsal thalamus and epithalamus are successively more dorsal parts of the diencephalic neural tube. In contrast, the neuromeric model considers the hypothalamus to represent the basal (ventral) part of the neural tube, associated with the dorsally located telencephalon (together representing the anterior forebrain or secondary prosencephalon; i.e., the most rostral piece of the neural tube). Accordingly, retromammillary region, posterior tuberculum (in the zebrafish corresponding to ventral and dorsal posterior tuberculum, respectively) and basal synencephalon represent three basal (ventral) forebrain portions associated with the corresponding alar plate portions (i.e., ventral thalamus, dorsal thalamus, pretectum) of the caudally adjoining piece of the neural tube (together forming the three prosomeres constituting the posterior forebrain, i.e., anterior parencephalon, posterior parencephalon and synencephalon). This example shows that it is far from trivial to disclose which CNS *bauplan* or model one uses to interpret the correct topological position of a particular area within the neural tube.

1.2.3. The Next Dimension: Understanding Central Nervous Proliferation and Neurogenesis/Gliogenesis

Along with the emergence of an increasingly adequate picture of relevant transverse and longitudinal elements of the vertebrate CNS, important concepts of cellular (neural) proliferation arose through the 20th century. In principle, the neural tube could increase its size and develop various vesicles and neuromeres described above without cell proliferation; indeed, cell shape changes and transitory increase in liquor pressure have been implied in these processes. However, there is now ample evidence that cell proliferation (followed by migration and differentiation) is indeed involved in the shaping of the early CNS.

Historically, the Swedish comparative embryological school initiated by Nils Holmgren (and continued by Palmgren, Bergquist and Källén) plays a crucial role in the understanding of how cellular proliferation relates to morphogenetic processes in the CNS, such as, for example, in the establishment of neuromeres. The central concept of the Swedish school is that the origin of adult brain structures from a particular site within the proliferative periventricular sheet of the neural tube is decisive for the final phenotypic appearance of those brain structures, and, particularly, for their comparative interpretation, that is, for identifying homologies among brain structures in different vertebrate brains. Clearly, at the core of this concept are local differences in the spatiotemporal proliferative behavior in the periventricular neural tube sheet and the subsequent migratory and differentiation behavior of brain cells originating there. This introduces the third dimension necessary to understand the complexity of brain morphogenesis in addition to those two discussed above (i.e., transverse and longitudinal neural tube elements), namely the developmental relationship of the proliferative ventricular zones (matrix) with the more peripheral (subpial) postmitotic cellular central nervous architecture deriving from these ventricular zones (fig. 2).

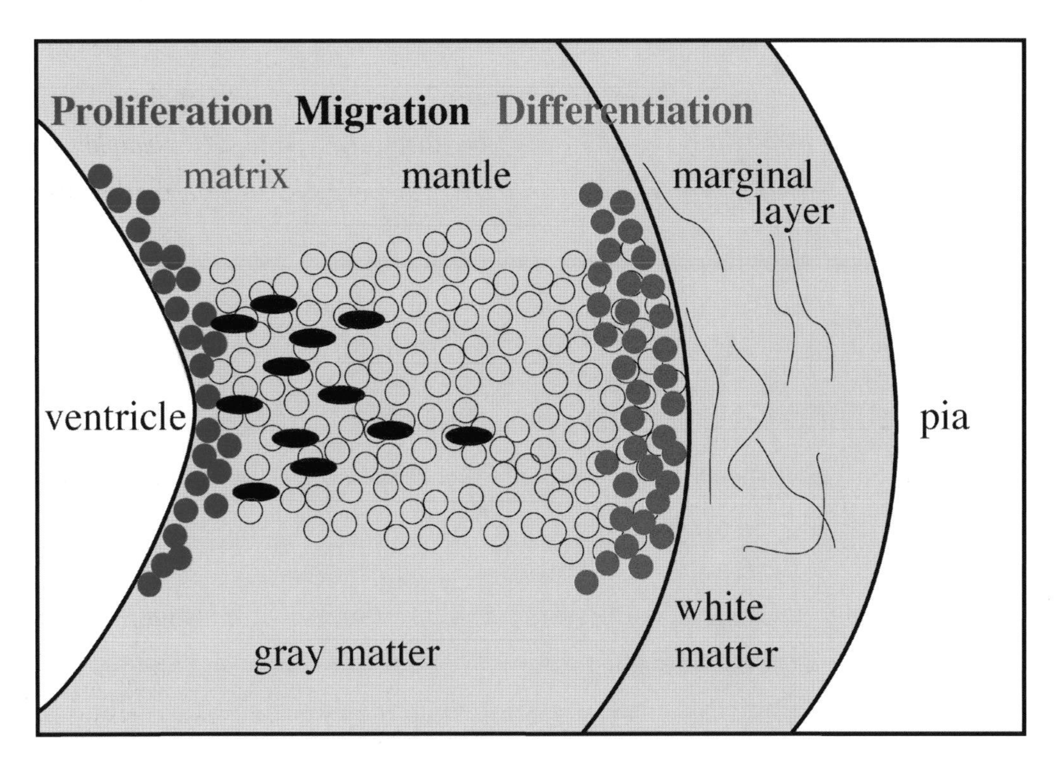

Fig. 2. Schematic transverse view of the developing vertebrate neural tube wall emphasizing distribution of proliferative, migrating and differentiating neural cells from ventricle to pia.

Using normal histology, Bergquist (1932, 1954) and Källén (1951, 1952; see also Bergquist and Källén 1954) brought this approach to fruition and described a grid of proliferating *matrix zones* (*Grundgebiete*) in the early vertebrate brain. These matrix zones were described to give off neurons in a locally specific spatiotemporal order, and—together with the subsequent migratory and differentiation behavior of fresh neurons leaving these matrix zones (*migration areas*)—to determine the final morphology of every brain region (fig. 3). This grid of matrix zones contained longitudinal as well as transverse zones and, ultimately, resulted in a neuromeric *bauplan* of the brain (see above). Since this work was comparative, the aspect of a common *bauplan* shared between all vertebrates was emphasized.

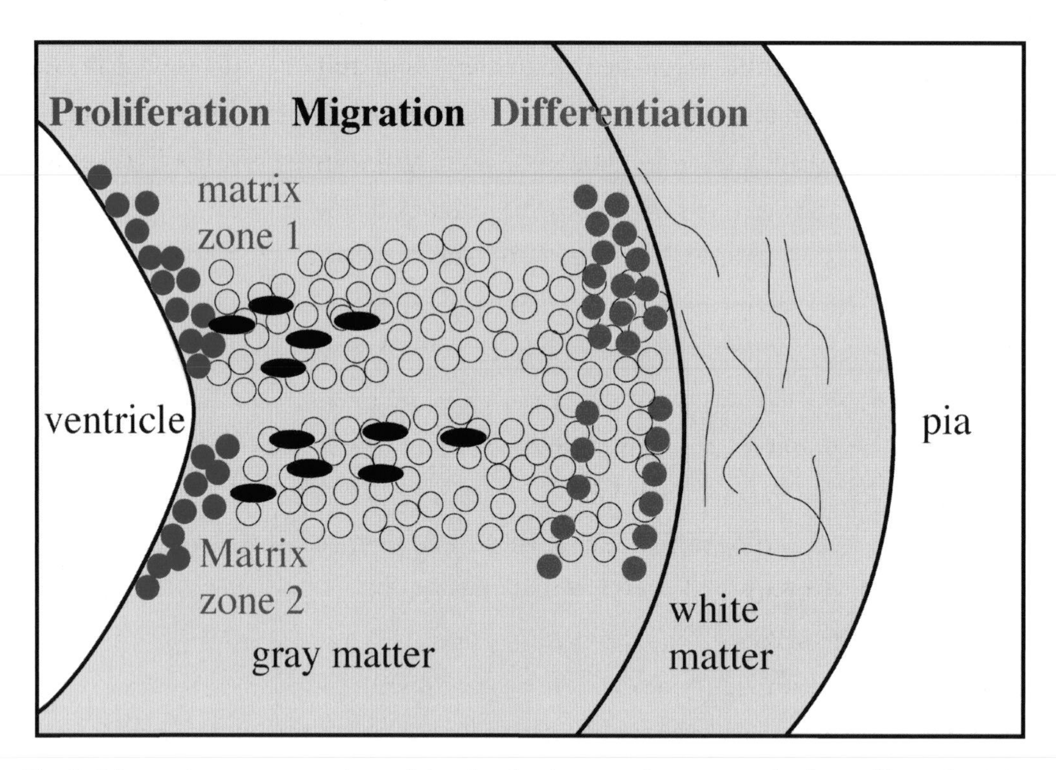

Fig. 3. Schematic transverse view of the developing vertebrate neural tube wall emphasizing local differences in neural proliferation and migration dynamics and their phenotypic results as viewed by Bergquist and Källén (see text).

As already mentioned in the context of neuromery, C.J. Herrick (and his followers Ariëns-Kappers, Huber and Crosby) vigorously attacked this concept which seemed to him only to please detail-loving morphologists, deviating from the functional longitudinal zones considered truly relevant for understanding brain structure and comparative issues. With regard to searching for homology of brain structures in different vertebrates, the American school founded the long-lasting approach of studying functional adult neuronal connectivity and neurobiologists—using an impressive series of ever more neuronal tracing substances—indeed apply this strategy extremely successfully to the present day. Although powerful in revealing neuronal connectivity and, thus, delivering arguments for homology of central nervous structures among vertebrates, this approach must fail in cases of similar circuitry that evolved in parallel (homoplastic evolution) in different species. Another contemporary, L. Edinger was also against the ideas of the Swedish school. He alternatively proposed the sequential addition of forebrain parts during vertebrate evolution which is incompatible with an ancestral common *bauplan*. Much in the vein of the Swedish school again was the "formanalytic approach" of H. Kuhlenbeck (publications between 1926–1973; see Nieuwenhuys (1998c)) who stressed that only the origin from a topologically comparable ventricular site, not adult circuitry, is decisive for a homology of brain structures, irrespective of their adult functional contexts.

What to make of this conflict today? Clearly, the Swedish school was at the disadvantage of lacking a powerful methodology (such as fate mapping) for substantiating their claims. However, many subsequent developmental research lines using cellular and molecular methods supported their concepts as we shall see below. Except for the interpretation of longitudinal zones in the forebrain (see above), the concept of a *bauplan* of matrix zones of the Swedish school was never really fundamentally incompatible with that of the functional zones of the American school, but rather complements it from the developmental point of view.

Later, autoradiography using tritiated thymidine (pioneer studies: Angevine and Sidman 1961; Fujita 1963, 1964, 1966; Bayer and Altman 1974; Rakic 1974; reviews: Bayer and Altman 1987, 1995a,b; Rakic 2002, 2003a,b) allowed to directly demonstrate the location of proliferating neural cells (short survival time after incubation) or, alternatively, the birth dates (i.e., time of last mitosis) and adult fates of neural tube cells (long survival times after incubation). These studies revealed that a *pseudostratified epithelium* exists in the early neural tube, which had been suggested already in the 1930s by Sauer (1935). During this neural tube stage, every neural cell is proliferative and has both a ventricular and a pial attachment. The cell nucleus migrates periodically from ventricle to pia (interkinetic nuclear migration), always going through mitosis and cell division close to the ventricle and DNA replication close to the pia. Later, the developing neural tube shows real layers, i.e., a ventricularly located proliferative cellular *matrix layer*, a more peripheral postmitotic cellular mantle layer, together forming the gray matter, and a marginal (largely non-cellular) white matter layer, consisting of first neurites of differentiated cells as well as of incoming extrinsic axons (compare with fig. 2). It also became clear that the matrix layer is not uniform along the neural tube ventricular surface, but rather shows local differences in proliferation and migration dynamics. According to Bayer and Altman (1995b), the early neural proliferative matrix zones represent a "blueprint (*bauplan*) of the anatomy of the CNS". Highly consistent with these views, the concept of *histogenetic units* (Puelles and Medina 2002) adds the element of the genetic pathways that control differently developing neural tube units characterized by a given matrix zone (some examples will be given in the final, comparative chapter).

How do freshly postmitotic neural tube cells born at the ventricle manage to migrate peripherally once the neural tube goes beyond the stage of the ubiquitously proliferative pseudostratified epithelium? Already Ramon y Cajal (1890), among other contemporaries, documented the cytology and histology of an early differentiated neural cell type, the *radial glia cells*, which exhibit a periventricularly located cell body and a long peripheral (radial) process extending towards the pial surface (fig. 4), therefore resembling somewhat the earlier proliferative neuroepithelial cells forming the pseudostratified epithelium. However, the role of radial glia in the guidance of peripherally migrating neural tube cells has been addressed only in the 1970s (Rakic 1971, 1972, 1974), and its dynamics and molecular basis are best documented in mammalian isocortex (neocortex) and cerebellum (Rakic 1988, 2002, 2003a,b; Hatten 1990, 1993, 1999, 2002). Apparently, both the action of diffusible repellents/attractants and of surface-mediated interactions with radial glia processes are closely intertwined in the process of peripheral migration of neural cells (Rakic 2003b). The latter have been mostly believed to originate from precursor cells different from radial glia cells.

Recently, the radial glia story has taken an unexpected and even more exciting turn. In addition to the meanwhile undisputed role of mammalian cortical radial glia as a guiding/scaffolding device for peripherally migrating neural cells, radial glia has been demonstrated to *produce* such migrating cells (Götz et al. 1998, 2002; Malatesta et al. 2000; Hartfuss et al. 2001; Noctor et al. 2001; Tamamaki et al. 2001). In the rodent cortex, differentiated radial glia cells—maintaining their peripheral radial process towards the pia—were discovered to remain proliferative and possess multipotency with regard to the generation of neurons and glia: they divide asymmetrically and give rise to most (i.e., projection) neurons in the cortical layers and, subsequently, to astroglial cells. This is not to exclude that there may be additional and/or more restricted neuronal and glial progenitor populations—possibly also deriving from radial glia cells—at certain developmental time points, especially in primates (Rakic 2003b). An evolutionary novelty only seen in the developing mammalian isocortex and basal ganglia is the *subventricular* (or subependymal) *zone* of proliferative cells. The early basal ganglia

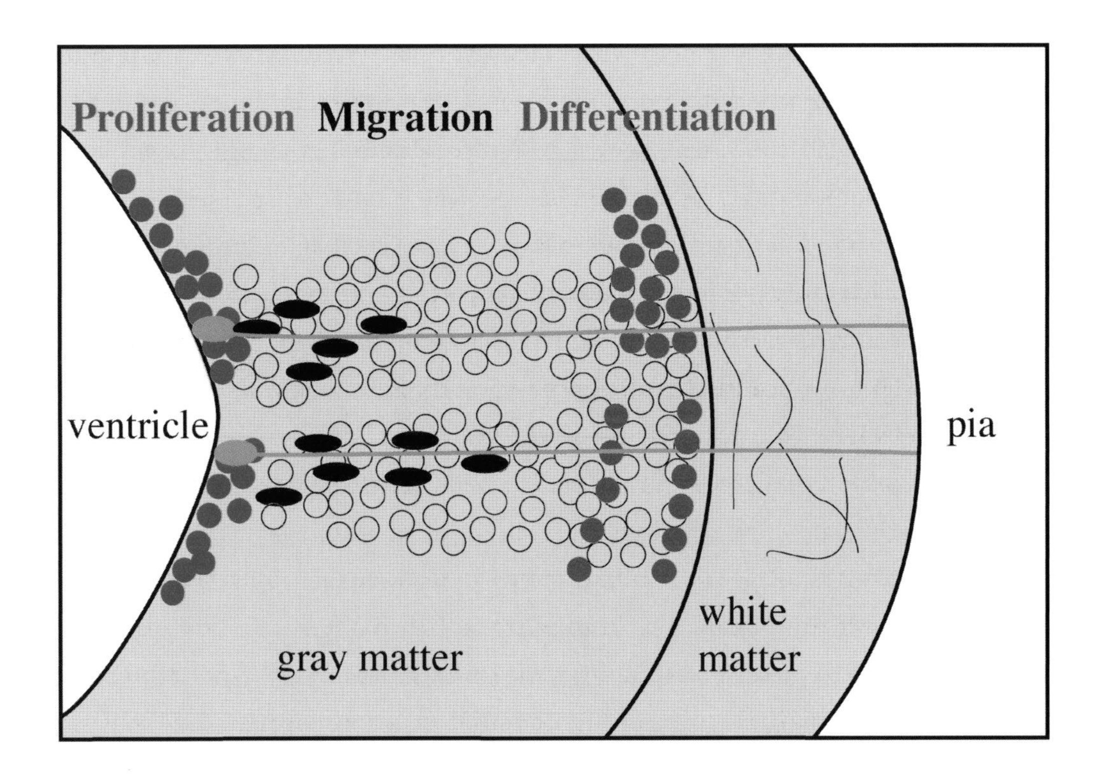

Neural Cell Type	Icon
proliferative neural cell	
radial glial cell (proliferative potential)	
cell with undefined state	
migrating determined neuronal cell	
differentiating neuronal cell	

Fig. 4. Schematic transverse view of the developing vertebrate neural tube wall emphasizing the role of radial glial cells as a guiding device for migrating neural cells (see text).

consist of the lateral and medial ganglionic eminences, LGE and MGE (with the LGE representing the future striatum and the MGE giving rise to the future pallidum), which lie lateroventrally to the cortex. The subventricular zone has been described to lie somewhat more remote from the ventricle than the proliferative ventricular zone and likely is an adaptation to the needs of large cell production (Smart 1976; reviewed in Voogd et al. 1998). Whereas radial glia cells are abundant in the subventricular zone of the isocortex (Smart et al. 2002; Malatesta et al. 2003), they are absent in the subventricular zone of the ganglionic eminences (Malatesta et al. 2003; see below).

Although radial migration of newborn neural cells along radial glia fibers has been studied mostly in the mammalian cortex and cerebellum, it may be considered prototypical for the entire developing vertebrate neural tube, because radial glia cells principally occur in every region of the vertebrate CNS, at least at early stages (Edwards et al. 1990; Hatten 1999). Furthermore, the multipotential role of radial glia in the generation of neural cells could be general for the entire CNS. This would be consistent with the fact that anamniotes exhibit a greater potential for adult neurogenesis than amniotes and retain radial glia in adult brain regions, whereas mammalian radial glia cells mostly do not persist into adulthood and are believed to at least partially transform into astrocytes (Götz et al. 2002; Rakic 2003a). However, the proportion of radial glia-derived neural cells varies greatly even within the mammalian telencephalon. In the ganglionic eminences, the radial glia (of the ventricular zone) generates only a small fraction of neural cells and an even smaller fraction of neurons, i.e., striatal and some long-distance migrating olfactory bulb interneurons (Malatesta et al. 2003). Most of these (granule and periglomerular) olfactory bulb cells, as well as the striatal projection neurons (see below) are produced by non-radial glia progenitors of the subventricular zone of the lateral ganglionic eminence (Campbell 2003; Stenman et al. 2003). Furthermore, the radial glia processes in the developing striatum lose their pial contact much earlier than in the developing isocortex (Smart 1985). Therefore, striatal radial glia guided cellular migration appears insignificant compared to passive aggregation of postmitotic cells, making the presence of differentiated radial glia cells obsolete much earlier than in the cortex. In the ganglionic eminences, a thick embryonic proliferative subventricular zone of non-radial glia cells is formed and it is the source of the overwhelming majority of striatal and pallidal projection neurons and interneurons (Malatesta et al. 2003), including those interneurons destined to invade the cortex (see below). However, the later developing, postnatal subventricular zone cells derive from radial glia cells and are gliogenic (Malatesta et al. 2003). Thus, the real question is how subpallial (and any other central nervous) glial and neuronal progenitors different from radial glia form out of the proliferative cells of the pseudostratified epithelium: do these progenitors originate from very early radial glia and lose their glial characteristics or, alternatively, do they derive even earlier from the neuroepithelial cells directly?

Although radial glia guided migration may be considered the predominant mechanism of cellular translocation in the developing vertebrate neural tube (Rakic 1988; Hatten 1999), important exceptions to this general rule do exist. There are several well-established examples of non-radial glia guided, so-called *tangential cellular migrations* during CNS development, because in these cases cells move perpendicular to the radial glia processes. For example, there is tangential migration of a minority of intrinsically generated (glutamatergic) isocortical projection neurons (O'Rourke et al. 1992), as well as long-distance tangential migration of GABAergic (destined for isocortex and striatum) and cholinergic (destined for striatum) interneurons originating in the subpallial medial ganglionic eminence (Marìn et al. 2000; see also discussion in the final, comparative chapter). The rhombic lip which forms the crest of the V-shaped medulla oblongata is the source of many tangentially migrating cells destined for basal brainstem structures, such as the inferior olive, the lateral reticular nucleus or the pontine formation (Bayer and Altman 1995b). The rhombic lip also gives off cells that invade the developing cerebellar plate and form the so-called external granular layer (EGL) lying directly below the pia (see section 1.3.3). Finally, there is a rostral migratory stream of neural cells into the olfactory bulb (maintained into adulthood) consisting of cells produced in the subventricular and ventricular zones of the dorsal part of the lateral ganglionic eminence which turn into granule and periglomerular cells (Luskin 1993; Wichterle et al. 2001; Campbell 2003; Malatesta et al. 2003; Stenman et al. 2003; see also above).

Somewhat unexpectedly, Altman and Bayer rejected both a strong role of radial glia guided migration and a close correspondence of their matrix zones to the *bauplan* of matrix zones suggested by Bergquist and Källén. But today, Altman and Bayer's results as outlined above appear highly compatible with the ideas described previously by the Swedish school (see above). Furthermore, radial (glia guided) cellular migration is clearly consistent with Bergquist and Källén's (1954) description of cellular migration waves originating in distinct neural proliferation matrix zones. The spatiotemporal regulation of these migration waves, plus the action of subsequent differentiation processes, leads to various differing morphologies along the neural tube axis—with the various contributions by tangentially migrating neurons being exceptions to the rule to be integrated into the general picture of morphogenesis. Today, evidence concerning the genetic control of these processes of neurogenesis (some of which will be discussed in the final, comparative chapter) is increasingly emerging, complementing the picture designed by earlier active neuroscientists.

After having laid the foundations to understand the most general principles of vertebrate CNS morphogenesis, we are now in a position to consider some cases of neural development of well investigated mammalian brain parts in more detail.

1.3. Functional Brain Architecture: Mammalian Examples

In the following, some examples of neuro- and gliogenesis in mammalian brain development shall be discussed because the data presented later in this atlas will directly shed some light on comparable events in early zebrafish brain development.

1.3.1. Cortex

The development of the mammalian cortex represents a special case of a matrix zone as conceived of in the foregoing. Cortex development does not primarily depend on many different, separate matrix zones with great local spatiotemporal differences in proliferation, migration and differentiation behavior as, for example, in the case of the dorsal thalamus where many different brain nuclei must be formed in a particular sequence (see below). The isocortex instead is of rather uniform thickness and histology (six layers, hence the name *isocortex*). Despite local differences in the emphasis of isocortical layers (layer 1 being most pial), they are unified by a common organizational theme: layer 4 is always the recipient of input (intracortical or external) and its cells project up into layer 2, the pyramidal cells of layers 2/3 form the intracortical (long-distance) efferents, whereas layers 5/6 give rise to extracortical efferents (for example, to thalamus or spinal cord). Layer 1 predominantly plays a developmental role.

The isocortex is initially formed by a huge continuous sheet of proliferative units and the crucial point is the maintenance of topological relationships between those proliferative units in order to allow the correct development of local specializations of future cortical areas (e.g., visual or motor areas), that is, positional information in the horizontal and vertical direction must be maintained.

According to the *radial unit hypothesis* of Rakic and coworkers (reviewed in Rakic 1988, 2002, 2003a,b), the entirety of proliferative units close to the cortical ventricle form a *protomap* of future cytoarchitectonically different cortical areas. The species-specific mosaic of adult cortical aeralization is the result of both intrinsic regionalization early within this protomap as well as of extrinsic influences acting on the maturing cortical neurons (such as the thalamic input in the sensory cortex). The descendant cells of these proliferative units maintain positional information in a peripheral position (forming a so-called radial column) because neurons originating in a common ventricular site share a common migratory pathway offered by the radial glia cells irrespective of sometimes convoluted relationship of ventricular site of origin and final subpial position. Additionally, the time of origin of a cortical neuron is decisive for its final position in the vertical dimension. Using the autoradiographic tritiated thymidine method, the generation of isocortical layers has been demonstrated to follow an *inside–out sequence* (Rakic 1974). With the exception of the very first born cells of the most peripheral layer 1 (e.g., Cajal-Retzius cells; Bayer and Altman 1995a; Price et al. 1997), early born cortical (projection) neurons come to lie close to the ventricle (i.e., layer 6) whereas later postmitotic neuronal cells migrate peripherally beyond them to populate the outer isocortical layers (with layers 2/3 being the last to form). In the Rhesus monkey, this process of isocortical neurogenesis takes about 60 days (roughly between embryonic day 40 and 100; total gestational period: 165 days), after which only glial cortical cells are formed (Rakic 1974).

Meanwhile, there is also good information about the genetic pathway involved in isocortical development. We shall discuss some of these data in the closing chapter comparing gene expression in different neurogenetic models after the presentation of various genetic and other markers visualizing various steps in neurogenesis throughout the zebrafish brain in the second chapter.

1.3.2. Thalamus

Development of the vertebrate thalamus has been studied using the autoradiographic tritiated thymidine method both in the rat (Altman and Bayer 1979, 1988; reviewed by Bayer and Altman 1995a,b) and mouse (Angevine 1970). The general similarity of concept and results on thalamic neurogenesis of these authors with those of the Swedish school (see above) regarding the presence of different matrix zones with differing proliferation, migration and differentiation behavior of descendant neurons has already been described above. In this part of the brain, spatiotemporal differences in the proliferation and migration behavior of neural cells of many separate matrix zones—and not of a continuous proliferative sheet as in the case of the isocortex—are of paramount importance for the generation of the adult mammalian thalamus. It represents one of the most complex nuclear assemblies in the vertebrate brain, subserving a multitude of functions in the relay of ascending sensory, limbic and arousal inputs to the cortex, as well as of cerebello-cortical motor loops. A critical thalamic structure is the reticular nucleus which ensheaths the dorsal thalamic nuclear masses laterally and acts inhibitory by way of GABAergic interneurons onto the dorsal thalamus. However, the origin of the reticular nucleus is from the ventral thalamic ventricular matrix zone.

It turned out that the thalamus represents a prime example of an *outside–in sequence* in neural tube development (as is also the case for the telencephalic septum and striatum, the hypothalamus, the preoptic area and the inferior colliculus; Bayer and Altman 1995b). In contrast to the isocortical inside–out developmental pattern of neuronal layers 2–6, the first born thalamic neurons migrate most peripherally towards a subpial position (for example, the laterally lying reticular nucleus). Subsequently born neurons destined to form the complex array of thalamic nuclei are generated in various separate dorsal thalamic matrix zones and their production is spatiotemporally choreographed as to come to lie in a increasingly less peripheral position, "below" the earlier born neurons with regard to the pial surface. Although radial glia clearly exists in the embryonic thalamus (Edwards et al. 1990), its role in the generation and migration guidance of neurons has not been addressed.

As in the case of the cortex, some information on the genetic pathways involved in thalamic development shall be discussed in the final comparative chapter, after various genetic and other markers visualizing various steps in neurogenesis throughout the zebrafish brain have been presented in Chapter 2.

1.3.3. Cerebellum

In mammalian (reviewed by Bayer and Altman 1995a,b; Hatten 1999) and avian (reviewed by Dubbeldam 1998) cerebellar development, different classes of neuronal cells are generated in subsequent waves. A first wave of postmitotic neuronal cells is born at the ventricle of the early cerebellar plate and it is destined to form the deep cerebellar nuclei, whose cells give rise to the efferent cerebellar output to other brain regions. Initially, these neuronal cells migrate somewhat towards the periphery, but then they move laterally down to the base of the cerebellum to their adult location below the cerebellar cortex. A second wave of postmitotic neuronal cells is then leaving the ventricle and moves towards the pial side of the cerebellar plate. These cells will eventually terminate their migration and become Purkinje cells, the only efferent cells of the cerebellar cortex projecting to the deep cerebellar nuclei. Additionally, the cerebellar cortex interneurons, that is the stellate cells of the most pially located molecular layer, and the basket and Golgi cells lying close to Purkinje cells, are believed to derive from the cerebellar plate ventricular layer (Goldowitz and Hamre 1998). A third wave of cerebellar neuronal cells is generated at the posterior boundary of the cerebellum towards the rhombencephalon, within the rhombic lip. The rhombic lip may be conceived of as the dorsal crest formed by the V-shaped walls of the medulla oblongata. The latter appears open dorsally, exhibiting the rhombic fossa, which is merely covered by a thin epithelial tela chorioidea. The rhombic lip remains highly proliferative for an extended time and it is the source of various hindbrain structures, finally located remote from the (dorsal) rhombic lip in a basal position, such as, for example, the inferior olive or the pontine cell masses (Bayer and Altman 1995b). Furthermore, neuronal progenitor cells from the rhombic lip directly caudally adjacent to the cerebellar plate migrate tangentially and rostrally into the cerebellar plate and populate its utmost periphery, directly below the pia. Since these neuronal progenitors will develop into cerebellar granular cells, this sheet of cells is called EGL. In contrast to the other classes of cerebellar cells described above, EGL cells remain mitotic for an extended period of time and give off an immense number of postmitotic cells which migrate basally (Komuro and Rakic 1998), using the radial glia specific for the cerebellum (Bergmann glia) as guiding devices during downward migration (Hatten 1993). These EGL cells will eventually cross the Purkinje cell layer to generate all of the densely populated, deep adult granular layer, which form the relay between various external inputs (mossy fibers) and the Purkinje cells (their second input being the climbing fibers from the inferior olive). It is believed that these migrating granular cells have a critical role in the proper arrangement (not the generation) of the Purkinje cell layer in that they stabilize the position of the Purkinje cells, a signal possibly mediated by the expression of *reelin* in the migrating granular cells (Curran and D'Arcangelo 1998). Thus, the EGL of the cerebellum represents a well-documented example for peripheral mitotic activity, i.e., subpial neurogenesis in the CNS, and one that is probably occurring in all gnathostome vertebrates.

1.4. Teleostean (Anamniote) Brain Development in Perspective

Putting zebrafish neurogenetics in perspective to amniote developmental neurobiology, a clear difference in emphasis becomes evident. The zebrafish has been chosen as anamniote model primarily for practical purposes of early accessibility of the embryo and faster genetic screening possibilities. Because of this reason, mostly the early zebrafish embryo and brain has been studied so far, while relatively little information is available on later zebrafish brain stages, which would be comparable to the late amniote embryo from which the knowledge on brain development and neurogenesis/gliogenesis stems that we have discussed above.

Thus, the zebrafish embryo, particularly around 24 h and earlier, represents the most intensively investigated life stage in this model animal. As other anamniotes, the embryonic zebrafish nervous system is dominated by *primary neurogenesis* which—implicitly or explicitly—is believed to involve the generation of mostly transitory neurons, such as somatosensory Rohon-Beard neurons (Bernhardt et al. 1990; Blader et al. 1997; Haddon et al. 1998; Reyes et al. 2004) and first spinal motor cells involved in reflex circuitry with the Rohon-Beard cells (Chitnis 1999), or the first differentiated brain neurons forming the early axonal scaffold (Chitnis and Kuwada 1990; Wilson et al. 1990; Park et al. 2000), although the transitory nature of the latter two classes of neurons has never been conclusively shown. At this point, one should note that in amniotes most of these early classes of neurons, such as the Rohon-Beard cells, are not developed and, thus, the term *primary neurogenesis* is used differently there. Hatten (1999) describes the major, radial glia guided production of neurons in amniote CNS development as primary neurogenesis, whereas amniote secondary neurogenesis often designates the later ongoing production and non-radial glia guided migration of neurons generated in *secondary matrix zones* (Bayer and Altman 1995b), such as the cerebellar EGL, the hippocampal dentate gyrus, and the ongoing neuronal production from the subventricular zone of the telencephalic lateral ventricle resulting in the rostral migratory stream to the olfactory bulb (see above). In contrast, in the zebrafish we consider as *secondary neurogenesis* the production of neurons outside of primary neurogenesis as defined for anamniotes above (and, thus, includes Hatten's primary and secondary neurogenesis in amniotes).

There is a pervasive utilization of whole mount in situ hybridization for visualizing and investigating gene expressions (and additional markers) and their effects in the developing early zebrafish brain. However, this level of resolution is unsatisfactory for understanding events comparable to those outlined above for the mammalian or bird brain with regard to the differentiation of the finer divisions of the CNS, such as the pallium (cortex), thalamus or cerebellum, which takes place mostly after 2 days of development in the zebrafish, i.e., during early postembryonic life (Kimmel 1993; Kimmel et al. 1995). These zebrafish brain stages shall be documented in atlas form in the Chapter 2 which forms the core of the present book.

1.5. The Approach of this Book

The focus of the present book is to go beyond (anamniote) primary neurogenesis and to present neuroanatomically detailed information about the local dynamics of neural proliferation, determination and differentiation, that is, the locally different spatiotemporal specificities of postembryonic zebrafish neural cell production, which forms the pillars of the adult brain architecture and which, logically, must be termed *secondary neurogenesis*. Once such information is available, it can be adequately related to those morphogenetic events described above for amniotes.

To this aim, we will visualize the activity of critical genes and some additional cellular markers—which are already known to play a role in primary neurogenesis—during postembryonic zebrafish brain development (i.e., secondary neurogenesis; see Chapter 2). Two related fields may be explored in this way. First, local conditions in brain neurogenesis are revealed (e.g., regarding maturation status or identification of locally acting genes). We will exemplify in the third chapter how the atlas information may be related to additional data and used in the analysis of neuro developmental questions in particular zebrafish brain areas. Equally important is that comparative aspects regarding the comparison of finer brain subdivisions with other model systems can be discussed more aptly than before with this atlas information at hand (as shown for a teleost–mammalian brain comparison in Wullimann and Mueller 2004b). We will present such a broad comparative analysis of the forebrains of neurogenetic model systems in the final chapter. Finally, it is our conviction as well as our hope that a fusion of these insights will allow to identify truly comparable functional CNSs among neurogenetic models and make the zebrafish more useful in neurobiological studies related to biomedical applications.

Atlas of Cellular Markers in Zebrafish Neurogenesis

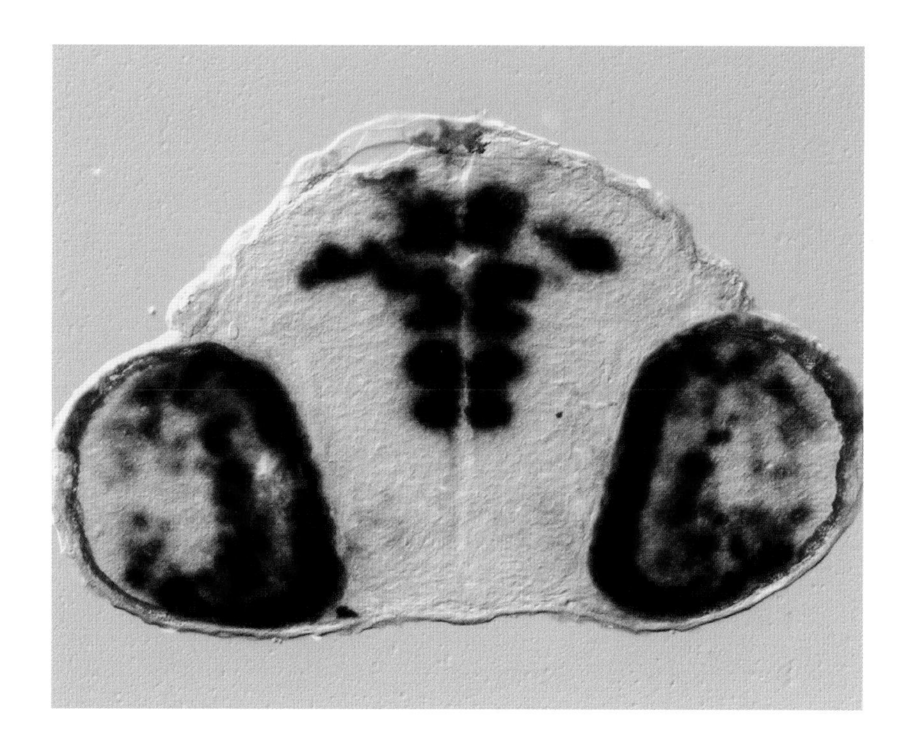

Chapter 2

Atlas

2. Atlas of Cellular Markers in Zebrafish Neurogenesis

The atlas pages presented in this chapter represent a greatly extended pictorial documentation of material previously treated in research reports (Mueller and Wullimann 2002a,b, 2003; Wullimann and Mueller 2002, 2004a; shortly summarized below). The purpose of the present chapter is to enable the readership to identify, at various postembryonic zebrafish stages, distinct neuroanatomical details in order to correlate those with information from other data sets and, thus, to facilitate neurogenetic analyses in the zebrafish.

2.1. Choice and Characterization of Molecular Markers

We turn now to the visualization of cellular/molecular markers (known to be involved in primary neurogenesis) during postembryonic, secondary neurogenesis in the zebrafish brain and the justification for their choice. We will apply a strong morphogenetic perspective established by "classical" neurobiological studies of vertebrate neurogenesis (treated in Chapter 1), i.e., a general framework of radial glia-guided ventriculopial migration of neuroblasts will be assumed. Therefore, certain expectations exist regarding the spatial distribution of the investigated molecular markers between ventricle and pial surface in the developing zebrafish neural tube: markers specific for proliferative, early determined and differentiating neuronal cells should be located in cells increasingly more remote from the ventricle in radial (pial) direction (see fig. 5).

Alternatively, the distribution of said markers might reflect local, special cases of tangential migration. As we shall see, both expectations are met by the data depending on the zebrafish brain region. First, the expected overall picture regarding radial glia-guided migration in neurogenesis (depicted in fig. 5) is identical to—and may thus be used at the same time as—a summary of findings on the actual distribution of markers for neurogenesis in most zebrafish brain areas. Secondly, there are also distinct cases where the distribution of markers is consistent with tangential migration, e.g., in the rhombic lip (i.e., the rim of the

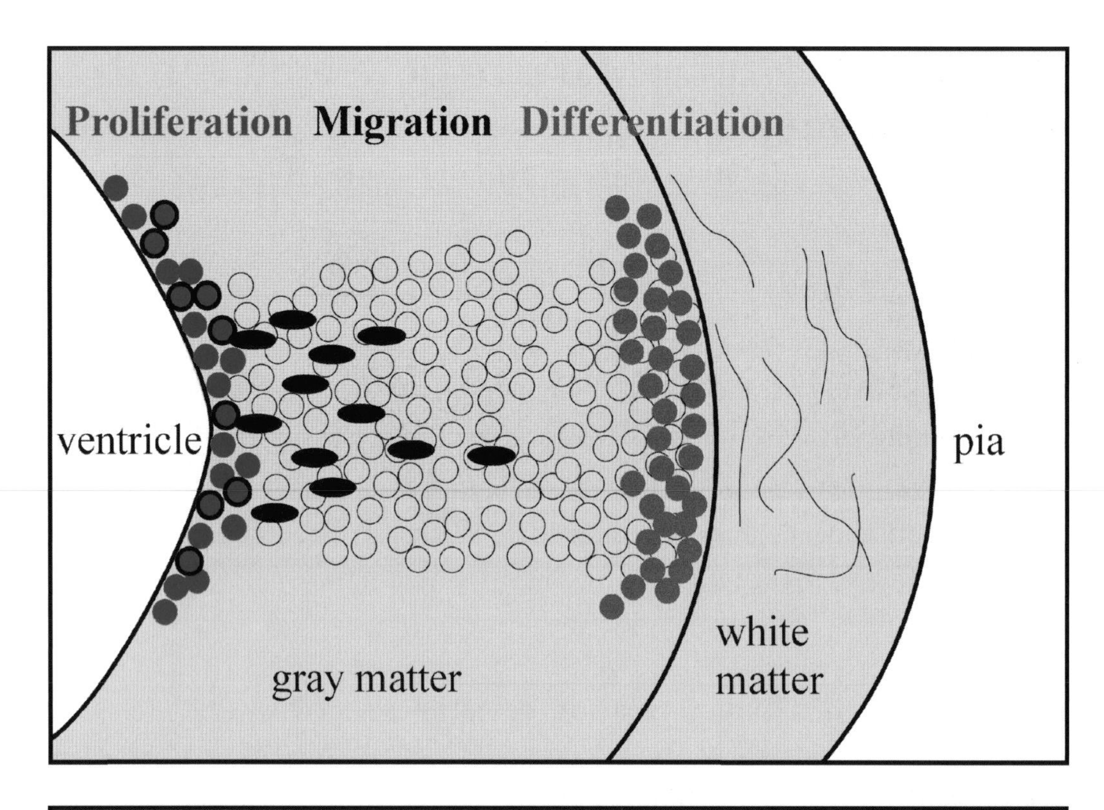

Neural Cell Type	Icon	Marker
proliferative neural cell	●	PCNA BrdU *notch1a*
cell with undefined state	○	
freshly determined neuronal cell	◉	*neurog1* *deltaA*
migrating determined neuronal cell	⬬	*neurod*
differentiating neuronal cell	●	Hu proteins

Fig. 5. Schematic transverse view of the developing vertebrate neural tube wall emphasizing again the distribution of proliferative, migrating and differentiating neural cells from ventricle to pia, but additionally including distribution of molecular markers shown in this chapter to be expressed during zebrafish secondary neurogenesis.

dorsal opening of the medulla oblongata, the rhombic fossa) (see also Chapter 3). We will also see that there are regions where certain molecular markers are mutually exclusively expressed, indicating local differences in genetic pathways in neurogenesis (see Chapter 3).

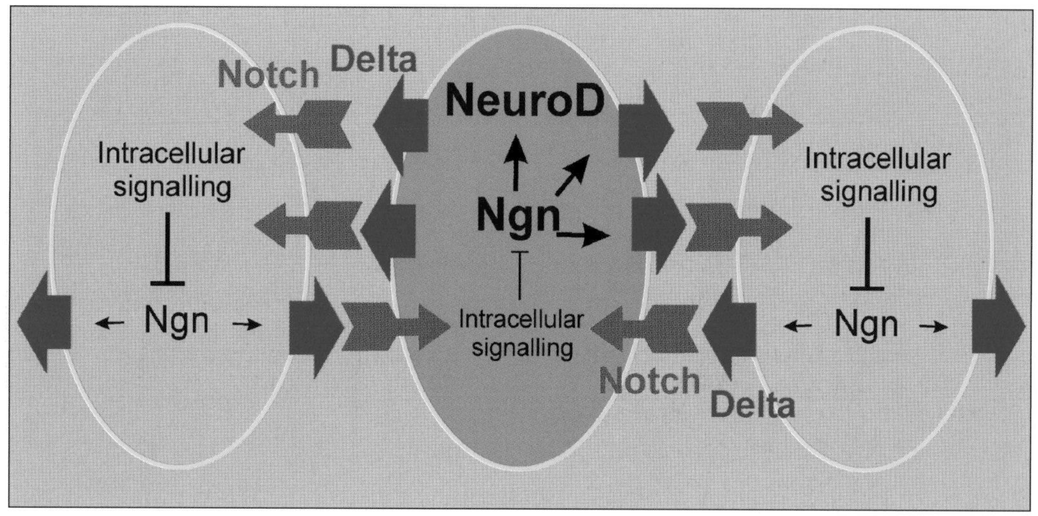

Fig. 6. Simplified schema of cellular and molecular interactions in the process of lateral inhibition and neuronal determination (ngn, neurogenin; see text for references and more details).

2.1.1. *Identification of Cellular Markers during Neurogenesis*

Pioneer studies in *Drosophila* have shown that various basic helix–loop–helix (bHLH) transcription factors (coded for by *achaete, scute, atonal* genes) are crucial for the establishment of neural competence as well as for the determination and differentiation of specific neural cell fates (briefly summarized in Mueller and Wullimann 2003). Subsequently, comparable findings regarding such "proneural" bHLH and additional genes emerged in anamniote vertebrate primary neurogenesis also. Thus, similar to processes in so-called proneural clusters of *Drosophila,* scattered cells in proliferative zones within three longitudinal stripes in the early anamniote vertebrate neural plate are singled out to become neurons by a mechanism of *lateral inhibition* (fig. 6). It is mediated by so-called "neurogenic genes", including the transmembrane protein coding genes *Notch* and *Delta;* the decision towards a specific neural cell fate is the result of interactions between proneural and neurogenic genes (Coffman et al. 1990, 1993; Chitnis et al. 1995; Dorsky et al. 1995, 1997; Chitnis and Kintner 1996; Lewis 1996, 1998; Ma et al. 1996; Haddon et al. 1998).

In short, the *atonal*-related proneural gene *X-ngnr-1* (*Xenopus neurogenin-related gene 1*) is expressed in these early neural plate stripes (Ma et al. 1996) and the bHLH transcription factor X-ngnr-1 (neurogenin1) plays a crucial role in the Delta–Notch-mediated mechanism of lateral inhibition. Thus, X-ngnr-1 positively influences the expression of *Delta*, while gene activity of *X-ngnr-1* is silenced by activated Notch protein. The early embryonic expression of *X-ngnr-1* both precedes and overlaps the expression of *X-Delta-1*, as expected for a neuronal determinative function of the former gene. Moreover, the expression of *X-ngnr-1* acts upstream of *XNeuroD*, which codes for a bHLH transcription factor promoting neuronal differentiation (Ma et al. 1996).

In zebrafish primary neurogenesis, the expression domains and roles of several zebrafish *Delta* and *Notch* orthologues and bHLH coding genes, i.e., *Zash1a* (*ascl1a*), *Zash1b* (*ascl1b*), *neurogenin1* (*neurog1*) and *neurod* were characterized earlier (Bierkamp and Campos-Ortega 1993; Allende and Weinberg 1994; Blader et al. 1997; Dornseifer et al. 1997; Westin and Lardelli 1997; Appel and Eisen 1998; Bally-Cuif et al. 1998; Haddon et al. 1998; Korzh et al. 1998; Appel et al. 2001); we reviewed this information briefly before (Mueller and Wullimann 2003). Going beyond these studies on primary neurogenesis in the zebrafish embryo, we have recently correlated the activity of those genes with specific neurogenetic processes during secondary neurogenesis in defined locations of the postembryonic zebrafish brain (Mueller and Wullimann 2002a,b, 2003; Wullimann and Mueller 2002, 2004a). These studies document the expression domains of several zebrafish bHLH genes (*neurod, neurog1, Zash1a, Zash1b*) and of *Notch* and *Delta* genes (*notch1a, notch1b, notch5, deltaA, deltaD*). We have explicitly related these gene expression data with the immunohistochemically determined distribution of cellular markers for proliferation, i.e., PCNA (*p*roliferating *c*ell *n*uclear *a*ntigen, the auxiliary protein of DNA polymerase δ; Wullimann and Puelles 1999; Wullimann and Knipp 2000) and BrdU (5-*br*omo-2′-*d*eoxy*u*ridine, a thymidine analogue incorporated into the DNA during the S-phase of mitosis, used to visualize proliferating cells; Mueller and Wullimann 2002b; see Chapter 3), and for neuronal differentiation, i.e., Hu-proteins (family of mRNA binding proteins with a gene-regulatory function at the posttranscriptional level known to be specifically expressed in neuronal cells; Marusich et al. 1994; Barami et al. 1995; Park et al. 2000; Mueller and Wullimann 2002b). Based on these previous studies, we choose to visualize here *notch1a, neurog1, neurod* expression domains, as well as PCNA (and BrdU, see also Chapter 3) and Hu-protein distribution to demonstrate in atlas form proliferative, early determined and differentiating neuronal cells in the postembryonic zebrafish brain. Information regarding *Zash1a* and *Zash1b* will be given in Chapter 3.

2.1.2. Major Expression Patterns in Neurogenesis between 2 and 5 Days

Compared to the embryonic expression of neurogenic (*notch, delta*) and bHLH (*neurogenin, neurod, Zash*) genes (Bierkamp and Campos-Ortega 1993; Allende and Weinberg 1994; Blader et al. 1997; Dornseifer et al. 1997; Westin and Lardelli 1997; Appel and Eisen 1998; Bally-Cuif et al. 1998; Haddon et al. 1998; Korzh et al. 1998), the distribution of these genetic markers is very different during early postembryonic life (starting around 2 days postfertilization, dpf). We found these genes and other markers differentially expressed in a highly ordered and stage-dependent manner between 2 and 5 dpf. The *zebrafish brain at 2 days* is unique in comparison to older postembryonic stages, in that key markers of neurogenesis are found to be expressed in a pattern expected for the beginning of massive overall secondary neurogenesis (see below), representing a first stage of zebrafish brain development comparable with what is described in amniote (especially mouse) brains in that context (see last chapter). Because the hatching period also starts at 2 dpf (Kimmel et al. 1995), the coinciding process of extensive secondary neurogenesis in the entire zebrafish CNS is highly useful for characterizing the beginning of the postembryonic stage of zebrafish brain development, although some secondary neurogenesis surely starts before 2 dpf.

Typically, distinct ventricular proliferation zones are seen in which lateral inhibition takes place mediated by neurogenic genes (*notch1a, deltaA*). The *notch1a* expression domains in the 2 dpf zebrafish brain deliver a qualitatively identical pattern of ventricular proliferation centers visualized by PCNA and BrdU immunohistochemistry at 5 dpf (Wullimann and Puelles 1999; Wullimann and Knipp 2000; Mueller and Wullimann 2002a,b; see fig. 7), indicating that a similar regionalization (including a prosomeric organization) of proliferation exists already at 2 dpf.

Thus, *notch1a* expression domains allow a description of proliferation zones in the entire zebrafish brain at the outset of postembryonic brain development not provided in such high resolution ever before. One example is the retina where—in contrast to the more ubiquitously distributed PCNA—*notch1a* expression is restricted to the peripheral edge, which was previously demonstrated to be the exclusive locus of retinal proliferation at 2 dpf (Marcus et al. 1999).

The *neurog1* and *neurod* genes are both expressed typically in a correlated and partially overlapping fashion in many—but not all—zebrafish brain regions,

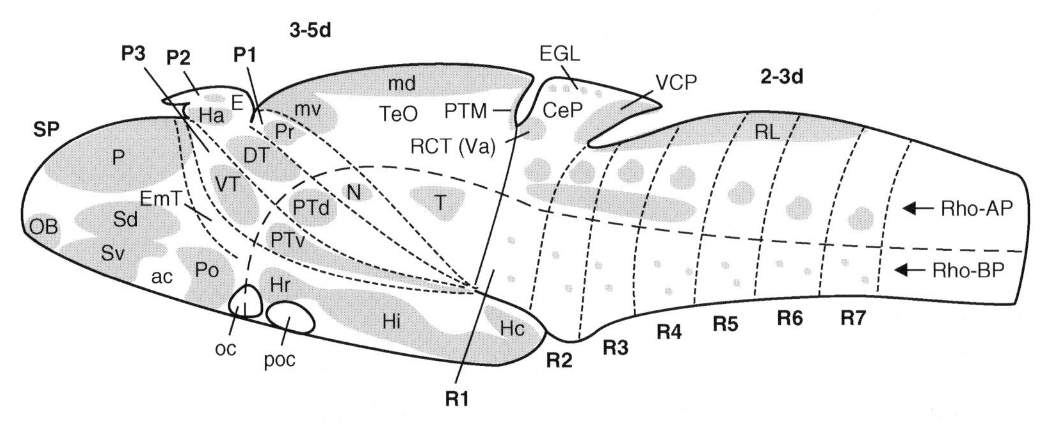

Fig. 7. Postembryonic zebrafish brain proliferation zones visualized either with PCNA (Wullimann and Puelles 1999; Wullimann and Knipp 2000) or BrdU (Mueller and Wullimann 2002b) interpreted within the prosomeric model (Puelles and Rubenstein 1993). Stippled line separates alar from basal plate proliferation zones and indicates forebrain anteroposterior axis bending. ac, anterior commissure; CeP, cerebellar plate; DT, dorsal thalamus (thalamus); E, epiphysis; EGL, external granular layer; EmT, eminentia thalami; H, hypothalamus; Ha, habenula; Hc, caudal hypothalamus; Hi, intermediate hypothalamus; Hr, rostral hypothalamus; md, mediodorsal tectal proliferation; MO, medulla oblongata; mv, medioventral tectal proliferation; N, region of the nucleus of medial longitudinal fascicle; OB, olfactory bulb; oc, optic chiasma; P1–P3, prosomeres 1–3; P, pallium; Po, preoptic region; poc, postoptic commissure; Pr, pretectum; PTd, dorsal part of posterior tuberculum; PTM, posterior tectal membrane; PTv, ventral part of posterior tuberculum; R1–R7, rhombomeres 1–7; RCT, rostral cerebellar thickening (valvula); RL, rhombic lip; S, subpallium; Sd, dorsal division of S; SP, secondary prosencephalon; Sv, ventral division of S; T, midbrain tegmentum; TeO, tectum opticum; Va, valvula cerebelli; VCP, ventral cerebellar proliferative layer; VT, ventral thalamus (prethalamus); Rho-AP, alar plate proliferation of rhombencephalon; Rho-BP, basal plate proliferation of rhombencephalon.

with *neurog1* being more restricted towards the ventricular proliferative zones and *neurod* extending more laterally. Comparisons with PCNA- and Hu-immunostains indicate that a great proportion of *neurog1*-positive cells are mitotic, but some appear to extend into the postmitotic gray matter where the *neurod* domains lie. Generally, expression of *neurog1* is restricted to those zebrafish brain regions where *neurod* is expressed (but see Chapter 3 for one problematic case). Expression of *neurod* is always at least one to several rows away from the ventricular surface and, thus, seems to flank ventricular proliferative PCNA and *notch1a*-expressing cells. However, the great majority of the more peripherally lying cells in the gray matter are *neurod* negative at 2 dpf. In contrast, Hu-positive neuronal cells are predominantly found at the lateral border of the gray matter, while the more medially lying cells are Hu negative. Forebrain clusters of *neurog1* and *neurod*-positive cells are arranged according to the prosomeric organization of the brain (similar to *notch1a* expression). It is important to note that subpallium, preoptic region, ventral thalamus and hypothalamus are clearly *neurog1* and *neurod* negative at 2 days.

In summary, a comparison of the distribution of these cellular markers in most parts of the 2 dpf zebrafish brain delivers a three-partitioned picture of the gray matter with ventricular proliferative (i.e., PCNA/*notch1a* positive) cells, freshly determined (i.e., *neurod* positive) neuronal cells which apparently migrate towards the periphery of the gray matter, and finally, cells (Hu positive) at the lateral border of the gray matter undergoing overt neuronal differentiation. The clear spatial separation of these cell masses in most brain areas likely represents three phases of neuronal development and is a unique feature of the 2 dpf zebrafish brain. In later postembryonic stages, Hu-positive cells extend down to the ventricular proliferating cells and are essentially complementary to these proliferative zones (Mueller and Wullimann 2002b). This difference of the 2 dpf brain compared to later stages may be due to the relatively high degree of proliferation dynamics and cellular migration in the 2 dpf zebrafish brain.

A comparison of the relative extents of PCNA (proliferative), *neurod* (freshly determined) and Hu-positive (differentiating) cell populations allows to determine the maturation state of a given brain part in the 2 dpf zebrafish brain. Some regions show a state of *delayed maturation* compared to other areas. In these immature regions, the above-mentioned three-partitioned distribution of PCNA-, *neurod*-, and Hu-positive cells from ventricle to pial surface is not obvious. For example, although in the mesencephalic tectum, anterior dorsal thalamus, entire cerebellar plate and rhombic lip region, the extent and shape of ventricular proliferation zones as revealed by *notch1a* expression are similarly restricted in the 2 dpf zebrafish brain as those described for later postembryonic (5 dpf) stages, there is a much greater extent of PCNA immunoreactivity (compare data in Mueller and Wullimann 2003), indicating that the proliferation and cellular migration dynamics are apparently much higher at 2 dpf. Accordingly, in the four regions just mentioned (tectum, dorsal thalamus, cerebellar plate, rhombic lip), expression of *neurod* (very recently determined neurons) is already strong, while Hu-positive (differentiating) cells are still largely missing.

The early valvula cerebelli (rostral cerebellar thickening)—where strong proliferative activity combines with absence of Hu-positivity **and** lack of *neurod* expression at 2dpf—is an even more exaggerated case of delayed maturation. This can be deduced from the fact that in the 3 dpf zebrafish brain, *neurod* is massively expressed in the valvula and Hu-positive cells are beginning to be present.

Thus, maturation of the valvula cerebelli at 2 dpf is even more delayed than in the examples given above (e.g., optic tectum, cerebellar plate, dorsal thalamus).

A problematic case regarding the otherwise strict correlation of *neurog1* and *neurod* expression exists in the cerebellum (cerebellar plate and valvula cerebelli) and lateral hinge-point region between cerebellar plate and medulla oblongata (which is where the eminentia granularis will develop), and the caudally adjacent rhombic lip. In all these closely associated regions *neurod* is massively expressed in early postembryonic stages (in the valvula later than in the other structures mentioned, see above), while *neurog1* expression is not found at all in the said structures between 2 and 5 dpf. It remains open, whether *neurod*-expressing cells are determined by *neurog1* in earlier stages (Korzh et al. 1998) or whether *neurod* expression is induced in this region by another bHLH transcription factor. However, the recently described zebrafish *neurogenin3* does not qualify for such a role because its expression domain is restricted to a very small portion of the zebrafish hypothalamus (Wang et al. 2001). Furthermore, our observations of bHLH gene expression from 2 to 5 dpf in the zebrafish cerebellar plate reveal a very special situation here, e.g., regarding the presence of the proliferative external granular layer in addition to a ventral proliferative layer and differences in gene expressions there, which will be treated in more detail in Chapter 3.

In the *3 dpf zebrafish brain*, the situation changes fundamentally regarding the above-mentioned three-partitioned picture of the gray matter exhibiting ventricular proliferative (i.e., PCNA/*notch1a* positive) cells, freshly determined (i.e., *neurod* positive) neuronal cells slightly more peripherally, and differentiating (Hu positive) cells at the lateral border of the gray matter. The Hu-positive cell masses are now essentially complementary to proliferative cells, leaving no gap between the two populations. Differentiation is now evident in the cerebellar region (valvula, cerebellar plate, eminentia granularis), which used to be Hu negative at 2 dpf. Thus, the cerebellar plate contains distinct Hu-positive cell populations, seemingly sandwiched between basal and peripheral layers of both PCNA-positive and *neurod*-expressing cells which are both seen in or close to ventral proliferative and external granular layers in the cerebellar plate. Also valvula cerebelli and eminentia granularis contain Hu-positive cells, but the (highly proliferative) rhombic lip remains free of Hu cells. However, some clearly postmitotic, differentiated areas remain (as an exception) Hu negative (e.g., retinal outer nuclear layer, subcommissural organ, part of the migrated posterior tubercular region M2, the Mauthner cells, rhombencephalic floor plate cells).

Brain proliferation zones at 3 dpf are nicely visualized with PCNA as being restricted to ventricular positions and they reveal qualitatively the same pattern as seen 1 day earlier with *notch1a* expression, including more restricted PCNA-positive zones in retina, optic tectum and cerebellar plate. The latter displays PCNA-positive cells in the ventral proliferative layer as well as in the external granular layer, which is now newly observable. There is also a newly recognizable, distinct PCNA domain in the eminentia granularis. The rhombic lip remains strongly PCNA positive. As at 2 dpf, patterns of proliferative zones reveal a prosomeric organization in the posterior forebrain (i.e., pretectum, dorsal thalamus, ventral thalamus) and rhombomeric organization in the medulla oblongata (compare with Wullimann and Knipp 2000). Small clusters of PCNA-positive cells are now present in the far migrated posterior tubercular region (M2, future preglomerular complex; see Chapter 3).

The *neurod* expression patterns at 3 dpf remain qualitatively similar to 2 days, although some domains start to thin out or are less extensive (e.g., in the basal synencephalon, i.e., the region of the nucleus of the medial longitudinal fascicle, in the midbrain tegmentum or in the anterior medulla oblongata). The migrated posterior tubercular region (future preglomerular region; M2) contains distinct, strongly *neurod*-positive cell clusters (see Chapter 3). The cerebellar plate now displays a basal and a peripheral (external granular layer) *neurod*-positive layer and the valvula cerebelli is additionally *neurod* positive (in contrast to 2 days). Also, the eminentia granularis is now clearly distinguishable by exhibiting a strong *neurod* signal. The rhombic lip appears largely *neurod* negative (in contrast to 2 days), but many *neurod*-expressing cells still emanate from it ventrally at the lateral periphery of the rhombencephalon. The subpallium, ventral thalamus, preoptic region and hypothalamus show complete absence of *neurod* expression also at 3 dpf.

In contrast to 3 dpf (where expression of *neurod* is qualitatively still rather similar to 2 dpf), the *5 dpf zebrafish brain* is characterized by a strong restriction of *neurod* expression (not shown in the atlas pages, but see Chapter 3) to a few particular regions (e.g., the two diencephalic migrated regions M1 and M2, or the cerebellar plate). Regarding proliferation, BrdU (5-bromo-2′-deoxyuridine), if used in a saturation label (see Chapter 3), reveals all central nervous proliferation zones and essentially corroborates the results gained previously with PCNA immunohistochemistry at 5 dpf (see above and fig. 7). Such brain proliferation zones typically lie directly at the ventricle along the entire neuraxis, flanked by a complementary mass of differentiating cells shown with Hu-proteins present in almost all the remaining gray matter, similar to 3 days. There are still sizable clusters of proliferative (BrdU-positive) cells in the far migrated pretectal region M1 and posterior tubercular region M2 (future preglomerular complex). The cerebellar plate still displays proliferative (BrdU-positive) cells in the medial and ventral proliferative zones, as well as in the external granular layer. There are also distinct BrdU populations in the eminentia granularis. The rhombic lip contains many BrdU-positive cells. In contrast, the remainder of the medulla oblongata shows few BrdU cells at all levels, confirming previous results with PCNA (Wullimann and Knipp 2000).

The situation regarding the distribution of Hu-positive cells at 5 dpf has not changed much compared to 3 days, as most gray matter cells are Hu positive and the ventricularly located proliferation zones remain Hu free (compare with BrdU). Similar to 3 days, regions with ongoing strong proliferation still show less Hu-positive cells compared with more mature regions. There is an increase in clearly postmitotic, differentiated areas compared to 3 days that remain Hu negative (e.g., retinal outer nuclear layer, subcommissural organ, lateral torus, diffuse nucleus of inferior lobe, cells in M2, Mauthner cells, rhombencephalic floor plate cells).

Compared with 3 days, the valvula cerebelli, the cerebellar plate and especially the eminentia granularis contain many more, strongly positive Hu cells, but the (proliferative) rhombic lip remains Hu free. However, in contrast to 3 days, many cells in the lateral medulla oblongata which appear to derive from the rhombic lip, are now strongly Hu positive.

In conclusion, these studies on secondary neurogenesis in the zebrafish represent the most detailed description of neurogenic/proneural gene expressions covering the entire brain in a vertebrate neurogenetic model system. The expression patterns confirm in detail the prosomeric organization delivered previously by proliferation studies (Wullimann and Puelles 1999; Mueller and Wullimann 2002b; see above). However, the particular bHLH genes investigated here represent only part of a larger group of orthologous/paralogous genes with potentially redundant functions. Moreover, transcriptional regulators visualized here and additional ones (bHLH and other factors) may be embedded in particular genetic pathways, which are only incompletely understood at present. Nevertheless, we think that our work will be helpful in future studies to fill in such information. The data presented here for the understanding of secondary neurogenesis in the zebrafish brain are likely to provide a basis to proceed at a faster pace in this model system than previously and, eventually, precede the detection of important neurogenetic processes in the zebrafish even in comparison to the mouse (see Chapter 3).

2.1.3. Some Regions Requiring Elevated Attention

The regionalization of the postembryonic zebrafish brain in this atlas is largely based on the initial description of zebrafish brain proliferation zones (Wullimann and Puelles 1999). The *ventral posterior tubercular proliferation* (PTv) extends into the part of the inferior lobe lying dorsal to the lateral recess ventricle (compare with fig. 7) and has been interpreted as the basal plate division of the third prosomere (ventral thalamus/prethalamus). This ventral posterior tubercular proliferation zone is caudally separated by a distinct white matter (corresponding in topology to the adult posterior tubercular commissure) from the proliferation zone of the caudal hypothalamus (Hc), which extends through the caudal remainder of the dorsal inferior lobe (and is considered here to be part of the basal plate of the secondary prosencephalon). The ventral part of the inferior lobe (lying ventral to lateral and posterior recess ventricles) contains the contiguous proliferation zones of intermediate and (ventral part of) caudal hypothalami. In the 2 and 3-day zebrafish inferior lobe, *neurod* expression is always clearly restricted to the dorsal inferior lobe as defined above, i.e., associated with the ventral posterior tubercular proliferation and *Zash1a* is restricted to the ventral inferior lobe (see Chapter 3). However, a close inspection at 3 days reveals some *neurod*-expressing cells posterior to the white matter separating ventral posterior tubercular and caudal hypothalamic proliferations. According to the initial definition, these cells would be located within the caudal hypothalamus (while the entire rest of the hypothalamus is free of *neurod* expression). Further investigation regarding origin and fate of these *neurod* cells in the caudal hypothalamus is therefore needed.

The *torus longitudinalis* is a highly conspicuous structure medial to the entire rostrocaudal extent of the adult optic tectal lobes and represents a special neural structure only seen in ray-finned fishes. While it cannot be discriminated as a separate structure in the 2 dpf zebrafish brain, we provisionally identify at 3 days a structure restricted to the most anterior optic tectum with strong *neurod* expression, but no PCNA positivity, as the postembryonic torus longitudinalis. This would mean that the torus longitudinalis develops along a rostrocaudal gradient and that it undergoes a considerable ventromedial displacement later. Also here, fate studies are needed to confirm this hypothesis.

There is a conspicuous, strongly *neurod* and Hu-positive region in the anteriormost region of the mesencephalic tectum where the latter curves ventrally around the tectal ventricle. We hypothetically identify this region as the *griseum tectale*, a retinorecipient structure described in birds (Garcia-Calero et al. 2002). If this is confirmed, the teleostean griseum tectale likely contributes to the adult superficial pretectal area, which would then have to be interpreted as partly derived from the mesencephalon, and not represent an entirely diencephalic/ pretectal region. However, this is problematic because a griseum tectale may be an avian specialization.

2.2. Technical Details

2.2.1. *A Word Regarding Terminology*

The terminology for postembryonic zebrafish brain structures stems primarily from three reports on the distribution of proliferative zones (Wullimann and Puelles 1999; Wullimann and Knipp 2000; slightly modified by Mueller and Wullimann 2002b; see fig. 7) which have revealed that postembryonic proliferation zones may be interpreted within a neuromeric context (Puelles and Rubenstein 1993). The studies mentioned have also shown that the degree of postembryonic brain differentiation (e.g., into nuclei or laminae) up to 5 days does not yet allow to apply neuroanatomical terms established for the adult zebrafish brain (Wullimann et al. 1996). Consequently, more general terms—widely accepted in comparative vertebrate neuroanatomy (such as pallium, subpallium, preoptic region, etc.)—have been used to describe postembryonic stages of the zebrafish brain and are applied in this atlas as well; many of these terms are also used for the embryonic brain (e.g., habenular, posterior, postoptic commissures, nucleus of medial longitudinal fascicle, etc.). In the revised neuromeric model (Puelles and Rubenstein 2003), dorsal and ventral thalami are designated as thalamus and prethalamus. We indicate this in all abbreviation lists, but keep the old terms in the figures in order to reduce confusion of the readership regarding the original literature on which this atlas is based.

Nevertheless, between the second and fifth day of zebrafish CNS development, progressing differentiation allows to recognize an increasing number of individual structures also present in the adult brain which shall be noted explicitly as we go along. The following atlas pages also include documentation of cranial nerve ganglia of the PNS; their identification follows that of Andermann et al. (2002), Liu et al. (2003) and Kerstetter et al. (2004).

The terminology for genes has been adapted to that of the ZFIN databank (Eugene, OR), although slightly different names have been used in our previous research reports.

2.2.2. Preparation of Zebrafish Brain Sections, Microscopy and Photoprocessing

For a detailed account on *technical details* (including protocols on immunohistochemistry and in situ hybridization), the reader may consult the original research papers (Wullimann and Puelles 1999; Wullimann and Knipp 2000; Mueller and Wullimann 2002a,b, 2003; Wullimann and Mueller 2002, 2004a). In short, zebrafish were kept and bred according to standard techniques (Westerfield 1995) and staged according to Kimmel et al. (1995). Zebrafish larvae between 2 and 5 days postfertilization used for in situ hybridization and immunohistochemistry were collected and fixed after anesthesia with tricaine methanesulfonate (MS 222; Sigma, Deisenhofen, Germany) or sacrificed after incubation with BrdU by shock-freezing at −70 °C.

For *immunohistochemistry* either Bouin's fixed (24 h fixation time of larvae), paraffin-embedded material (Hu-proteins, PCNA), cut at 7–10 μm, or freshly frozen sections (10–12 μm) of previously BrdU-incubated larvae (see Chapter 3 for more details), postfixed with methanol (20 min), were used. Specifications regarding primary antibodies are given in explanations facing the atlas pages. Details regarding immunostaining for Pax6 protein (one photomicrograph), are given in Wullimann and Rink (2001).

For *in situ hybridization* paraformaldehyde-fixed (overnight), paraffin-embedded material (section thickness: 7–10 μm) was used to visualize the expression of *neurog1*, *neurod*, *notch1a*, *Zash1a*, according to the protocol of Dorsky et al. (1995). Plasmids were kindly provided by Uwe Strähle and Patrick Blader (IGBMC-184-ULP; Strasbourg, France; *neurog1*, *neurod*), Eric Weinberg

(Philadelphia, PA; *Zash1a*), and Michael Lardelli (Adelaide, Australia; *notch1a*). Specifications regarding genes investigated are given in explanations facing the atlas pages.

For *microscopy* and *photoprocessing* a Zeiss Axiophot and a Zeiss Axiocam digital camera were used to analyze and photograph light microscopical sections in Nomarski optics; fluorescent sections were analyzed with a laser scanning confocal fluorescence microscope (Zeiss LSM 410 invert). Selected sections were slightly adjusted for contrast and sharpness using either Adobe Photoshop or Corel PHOTO-PAINT before mounting and labeled using CorelDRAW which was also used for preparing all remaining figures.

2.3. The End of Embryonic Life (2 Days): Outset of Secondary Neurogenesis

Atlas pages: *notch1a*

Atlas pages: *neurog1*

Atlas pages: *neurod*

Atlas pages: Hu-proteins

2.4. The Early Larva (3 Days): The Emergence of Brain Subdivisions

Atlas pages: PCNA

Atlas pages: *neurod*

Atlas pages: Hu-proteins

2.5. The Late Larva (5 Days): Increasing Differentiation

Atlas pages: BrdU

Atlas pages: Hu-proteins

notch1a (ZFIN ID: ZDB-GENE-990415-173; previous names: notch)
Description: neurogenic gene, codes for transmembrane receptor which binds transcellular ligand, expressed in proliferative (prospective) neural and sensory cells (and presomitic mesoderm in caudal embryo).

Gene Expression Domains in the 2-Day Zebrafish Brain

General appearance: Consistent with a role in lateral inhibition and singling out of neural/sensory cells, *notch1a* expression domains lie in ventricularly located proliferation zones of CNS. Expression patterns mimic those revealed with proliferation markers (e.g., prosomeric organization in posterior forebrain, i.e., pretectum, dorsal thalamus, ventral thalamus and rhombomeric organization in medulla oblongata), covering all known proliferative zones, including those in peripheral nervous system.

Description of levels on facing page

Sensory organs: Strong *notch1a* signal is present in retinal peripheral edge (which remains proliferative long after 2 days). Clusters of *notch1a*-positive cells are seen in olfactory epithelium, likely overlapping with proliferating cells.

CNS: In the telencephalon, pallial and subpallial *notch1a* domains, including the domain in medial proliferative zone of olfactory bulb, lie directly at the ventricular surface at all anteroposterior levels; this also applies to eminentia thalami and preoptic region. The latter is separated from the anteriorly lying subpallium by the anterior commissure. In the diencephalon, large blobs of expression are present in habenula, dorsal thalamus proper and ventral thalamus. All expression domains on facing page are interpreted as alar plate of anterior forebrain (secondary prosencephalon).

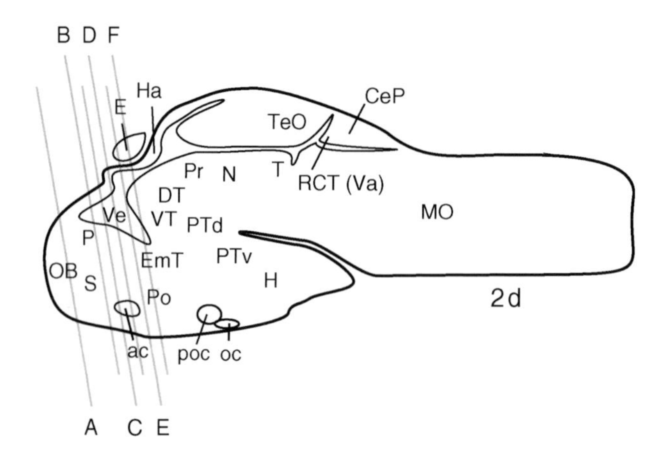

Abbreviations

ac	anterior commissure	P	pallium
CeP	cerebellar plate	Po	preoptic region
DT	dorsal thalamus (thalamus)	Poc	postoptic commissure
E	epiphysis	Pr	pretectum
EmT	eminentia thalami	PTd	dorsal part of posterior tuberculum
EPi	eye pigment	PTv	ventral part of posterior tuberculum
H	hypothalamus	RCT	rostral cerebellar thickening (valvula)
Ha	habenula	S	subpallium
lfb	lateral forebrain bundle	T	midbrain tegmentum
MO	medulla oblongata	TeO	tectum opticum
N	region of the nucleus of the medial longitudinal fascicle	Va	valvula cerebelli
OB	olfactory bulb	Ve	brain ventricle
oc	optic chiasma	VT	ventral thalamus (prethalamus)
OE	olfactory epithelium		

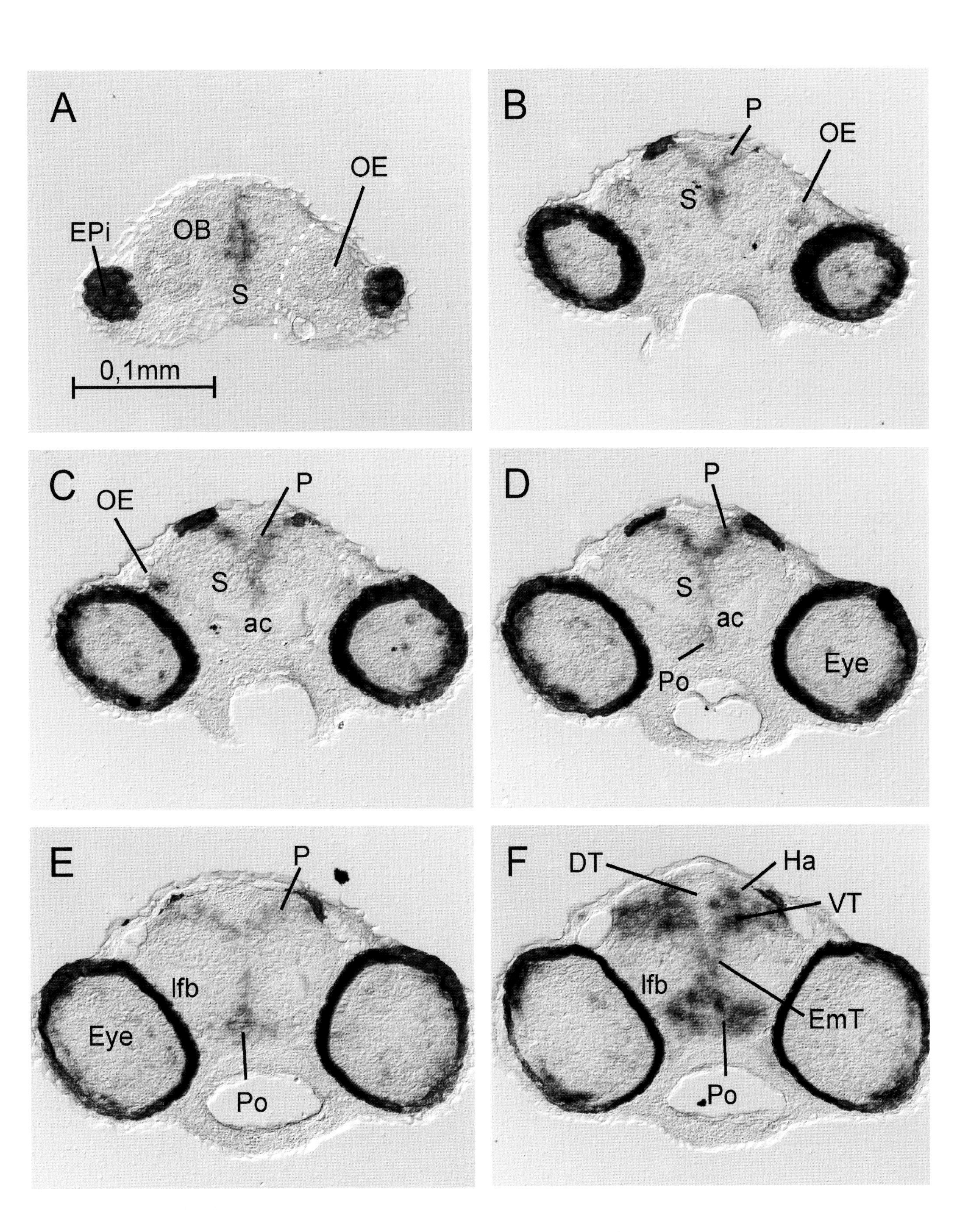
A
EPi
OB
OE
S
0,1mm
B
P
OE
S
C
OE
P
S
ac
D
P
S
ac
Po
Eye
E
P
lfb
Eye
Po
F
DT
Ha
VT
lfb
EmT
Po

notch1a (continued, 2nd plate)

Gene Expression Domains in the 2-Day Zebrafish Brain

Description of levels on facing page

Sensory organs: Strong *notch1a* signal is present in retinal peripheral edge (which remains proliferative long after 2 days). Additional small patches of *notch1a* expression are seen in central retina.

CNS: At these levels, a strong and extensive *notch1a* signal is present in the preoptic region, which is separated caudally by the postoptic commissure from the anterior part of the emerging hypothalamus. This rostral, as well as the intermediate hypothalamus (the latter is characterized by the lateral ventricular recess), display a strong *notch1a* signal. The hypothalamus is interpreted as basal plate of the anterior forebrain (secondary prosencephalon). In the diencephalon (posterior forebrain), large blobs of *notch1a* expression are present in habenula, dorsal thalamus proper and pretectum (interpreted as alar plate), as well as in the posterior tuberculum (dorsal and ventral divisions visible on facing page; interpreted as basal plate). Within the optic tectum (alar plate), large medial (dorsal and ventral parts visible rostrally) and lateral *notch1a* domains emerge. Except for a small ventricular spot, there is no *notch1a* expression in the zona limitans intrathalamica and in the epiphysis.

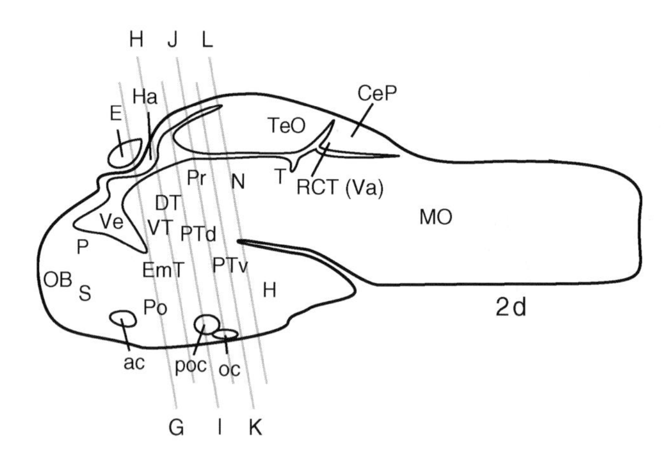

Abbreviations

ac	anterior commissure	oc	optic chiasma
CeP	cerebellar plate	P	pallium
DT	dorsal thalamus (thalamus)	pc	posterior commissure
E	epiphysis	Po	preoptic region
EmT	eminentia thalami	poc	postoptic commissure
H	hypothalamus	Pr	pretectum
Ha	habenula	PT	posterior tuberculum
Hi	intermediate hypothalamus	PTd	dorsal part of posterior tuberculum
Hr	rostral hypothalamus	PTv	ventral part of posterior tuberculum
l	lateral tectal proliferation zone		
LVe	lateral recess ventricle of H	RCT	rostral cerebellar thickening (valvula)
lfb	lateral forebrain bundle		
m	medial tectal proliferation zone	S	subpallium
md	dorsal part of m	T	midbrain tegmentum
MO	medulla oblongata	TeO	tectum opticum
mv	ventral part of m	Va	valvula
N	region of the nucleus of the medial longitudinal fascicle	Ve	brain ventricle
		VT	ventral thalamus (prethalamus)
OB	olfactory bulb	ZLI	zona limitans intrathalamica

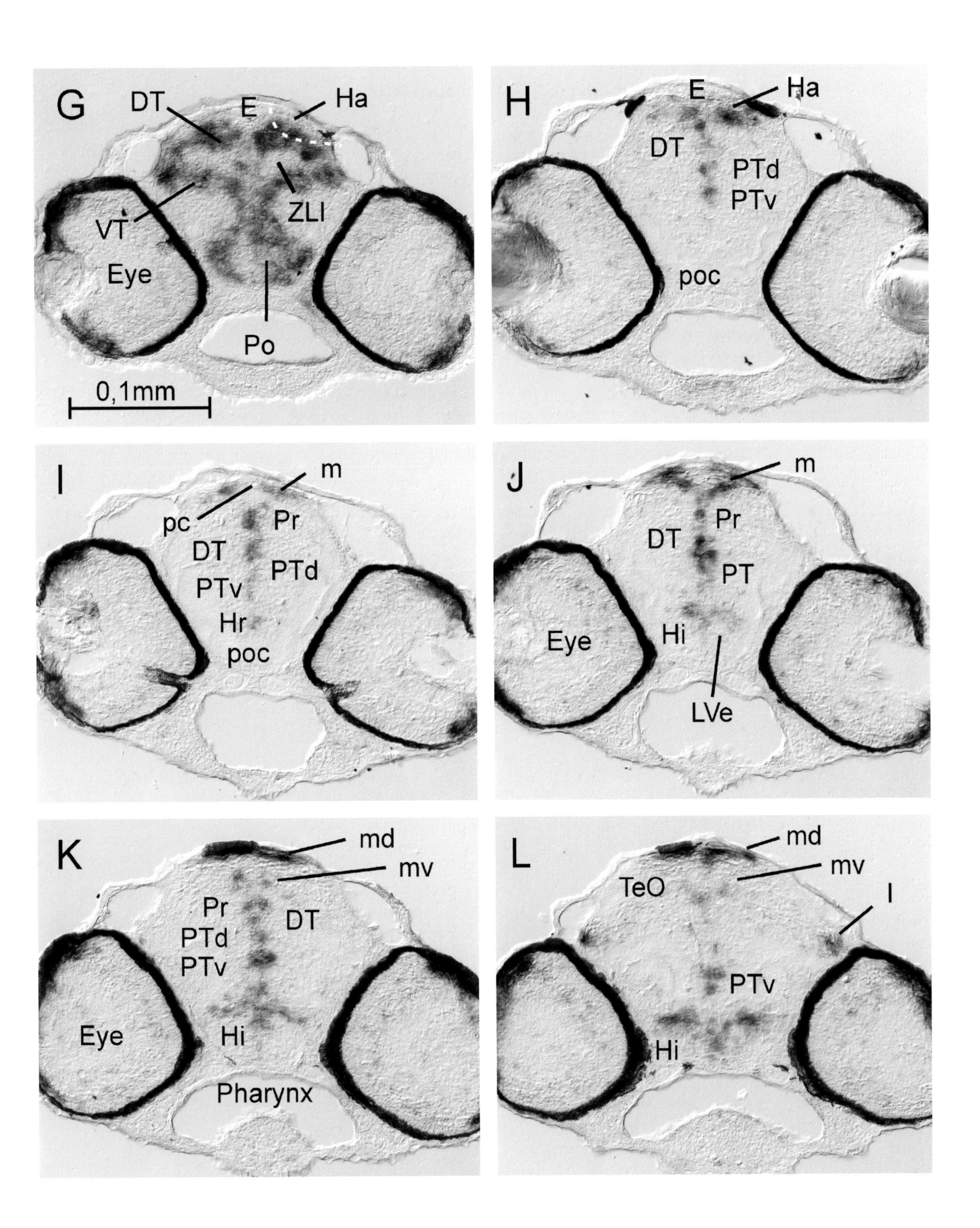
G
DT
E
Ha
VT
ZLI
Eye
Po
0,1mm
H
E
Ha
DT
PTd
PTv
poc
I
m
pc
Pr
DT
PTd
PTv
Hr
poc
J
m
Pr
DT
PT
Eye
Hi
LVe
K
md
mv
Pr
DT
PTd
PTv
Eye
Hi
Pharynx
L
md
mv
TeO
I
PTv
Hi

notch1a (continued, 3rd plate)

Gene Expression Domains in the 2-Day Zebrafish Brain

Description of levels on facing page

Sensory organs/PNS: Strong *notch1a* signal is present in retinal peripheral edge (which remains proliferative long after 2 days). Additional small patches of *notch1a* expression are seen in central retina. Expression of *notch1a* is also present in various cranial nerve ganglia of the peripheral nervous system.

CNS: Strong *notch1a* signal is present in intermediate and caudal hypothalami (the latter is characterized by the posterior ventricular recess). The hypothalamus extends into the ventral part of the inferior lobe and is interpreted as basal plate of anterior forebrain (secondary prosencephalon). The dorsal portion of the inferior lobe (interpreted as basal plate of ventral thalamic prosomere/P3) is *notch1a* negative at this level (but see previous page). Hypophysis shows weak and patchy *notch1a* signal. A small *notch1a* expression domain is seen in the synencephalic region of the nucleus of the medial longitudinal fascicle (N; basal plate of P1), which is replaced more caudally by distinct *notch1a* domain of midbrain tegmentum. Medial, basal (note that *notch1a* cells designated as b may partly lie ventral to tectal ventricle and, thus, belong to medulla oblongata) and lateral *notch1a* domains of optic tectum increase in extent caudally. Also the (medullary) midbrain–hindbrain boundary, and especially the most caudal optic tectum and rostral cerebellar thickening (future valvula), shows a strong *notch1a* signal. Most caudal midbrain also exhibits large *notch1a* domain in torus semicircularis. Anterior medulla oblongata contains *notch1a* domain along the ventral midline ventricle.

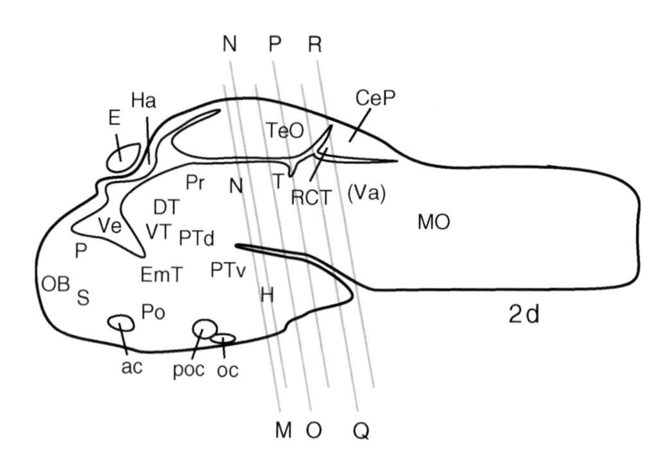

Abbreviations

ac	anterior commissure	OB	olfactory bulb
b	basal optic tectum	OC	otic capsule
CeP	cerebellar plate	oc	optic chiasma
Ch	chorda dorsalis	OG	octaval ganglion
DT	dorsal thalamus (thalamus)	P	pallium
E	epiphysis	Po	preoptic region
EmT	eminentia thalami	poc	postoptic commissure
EPi	eye pigment	Pr	pretectum
H	hypothalamus	PTd	dorsal part of posterior tuberculum
Ha	habenula	PTv	ventral part of posterior tuberculum
Hc	caudal hypothalamus		
Hi	intermediate hypothalamus	PVe	posterior ventricular recess of H
Hy	hypophysis (pituitary)	RCT	rostral cerebellar thickening (valvula)
l	lateral optic tectum		
m	medial optic tectum	S	subpallium
md	dorsal part of m	T	midbrain tegmentum
MHB	midbrain–hindbrain boundary	TeO	tectum opticum
MO	medulla oblongata	TS	torus semicircularis
mv	ventral part of m	Va	valvula cerebelli
N	region of the nucleus of the medial longitudinal fascicle	Ve	brain ventricle
		VT	ventral thalamus (prethalamus)

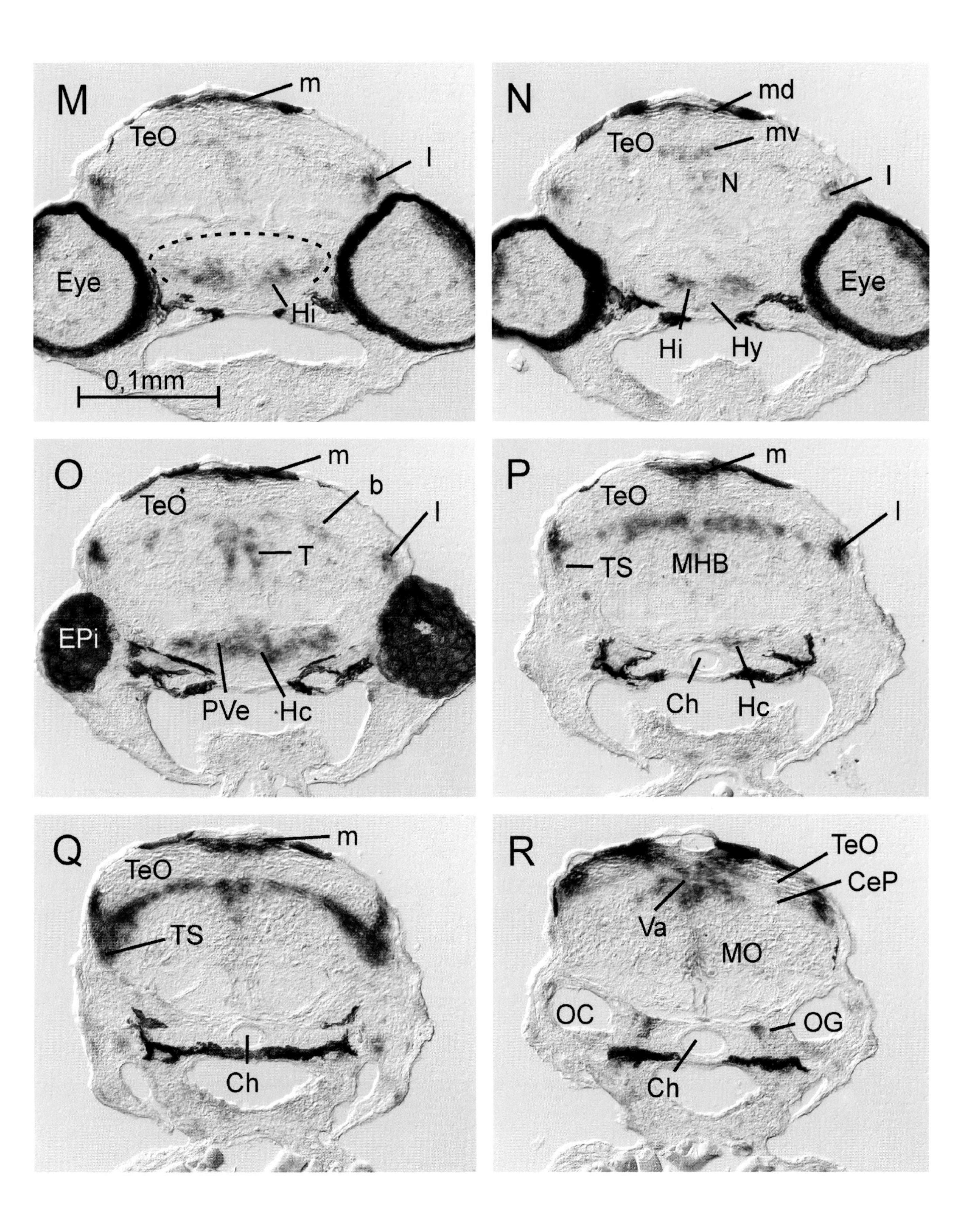
M
m
TeO
l
Eye
Hi
0,1mm
N
md
TeO
mv
N
l
Eye
Hi
Hy
O
m
b
TeO
l
T
EPi
PVe
Hc
P
m
TeO
l
TS
MHB
Ch
Hc
Q
m
TeO
TS
Ch
R
TeO
CeP
Va
MO
OC
OG
Ch

notch1a (continued, 4th plate)

Gene Expression Domains in 2-Day Zebrafish Brain

Description of levels on facing page

Sensory organs/PNS: Patchy *notch1a* signal in cellular lining of otic capsule and in various cranial nerve ganglia of peripheral nervous system.

CNS: At these levels, a series of globular *notch1a* expression domains is seen in basal proliferative layer of cerebellar plate (but not in proliferative external granular layer). At rostral to intermediate medullary levels (the latter is characterized by otic capsule), the rhombic lip (rim of dorsal opening of medulla oblongata, the rhombic fossa) is strongly *notch1a* positive. Approaching the obex (boundary to spinal cord), the rhombic lip fuses medially, closing the rhombic fossa. The medulla oblongata continues at these levels to be *notch1a* positive along the ventral midline ventricle and along the ventricular opening towards the rhombic lip. Some *notch1a*-expressing cells emanate some distance away from the rhombic lip ventrally into the more peripheral rhombencephalic gray matter.

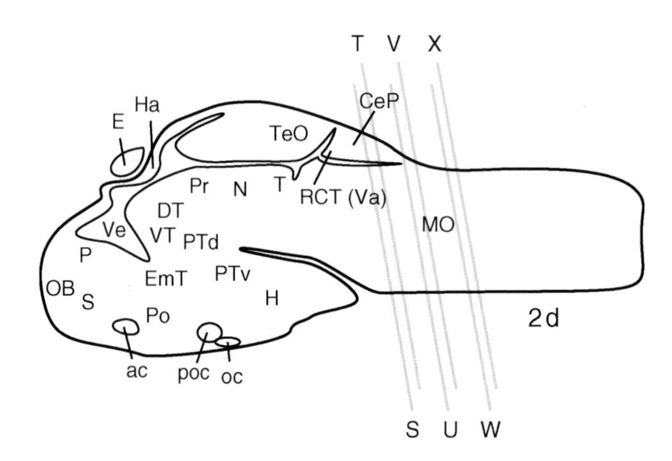

Abbreviations

ac	anterior commissure
CeP	cerebellar plate
Ch	chorda dorsalis
DT	dorsal thalamus (thalamus)
E	epiphysis
EmT	eminentia thalami
H	hypothalamus
Ha	habenula
MO	medulla oblongata
N	region of the nucleus of the medial longitudinal fascicle
OB	olfactory bulb
OC	otic capsule
oc	optic chiasma
OG	octaval ganglion
P	pallium
PLLG	posterior lateral line ganglion
Po	preoptic region
Poc	postoptic commissure
Pr	pretectum
PTd	dorsal part of posterior tuberculum
PTv	ventral part of posterior tuberculum
RCT	rostral cerebellar thickening (valvula)
RL	rhombic lip
S	subpallium
T	midbrain tegmentum
TeO	tectum opticum
Va	valvula cerebelli
Ve	brain ventricle
VG	vagal ganglion
VT	ventral thalamus (prethalamus)

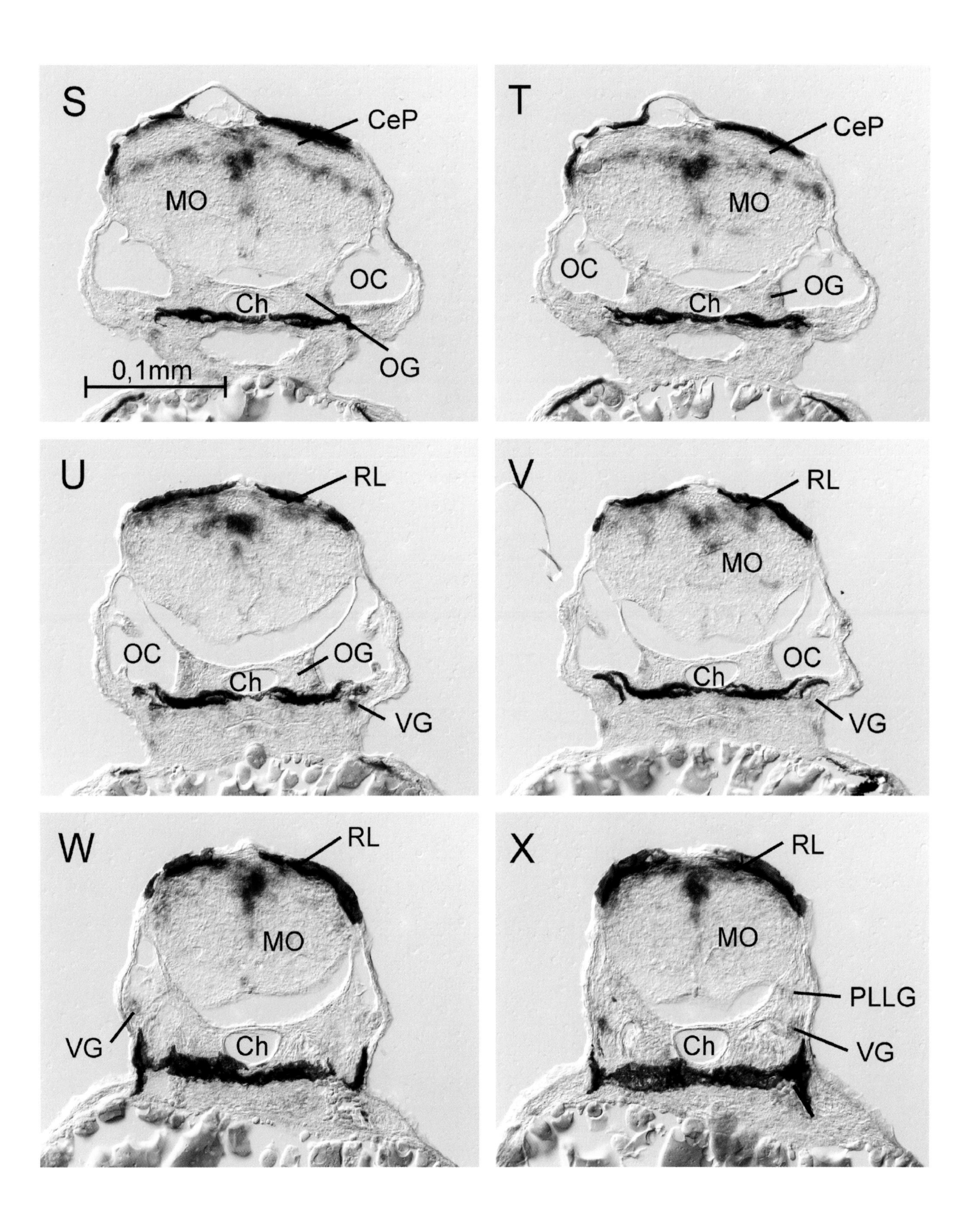
S
CeP
MO
OC
Ch
OG
0,1mm
T
CeP
MO
OC
OG
Ch
U
RL
OC
OG
Ch
VG
V
RL
MO
OC
Ch
VG
W
RL
MO
VG
Ch
X
RL
MO
PLLG
Ch
VG

neurog1 (ZFIN ID: ZDB-GENE-990415-174; previous names: ngr1; neurogenin; cb260; chun6899; zNgn1; ngn1; neurogenin1; neurod3)
Description: Proneural gene, codes for basic Helix–Loop–Helix transcription factor, expressed in early determined neuronal and sensory cells.

Gene Expression Domains in the 2-Day Zebrafish Brain

General appearance: The *neurog1* expression domains are closely parallel and partially overlap with the zones revealed with proliferation markers (e.g., prosomeric organization in posterior forebrain, i.e., pretectum, dorsal thalamus, ventral thalamus and rhombomeric organization in medulla oblongata), including PNS, which is consistent with a role in early neuronal determination upstream of *neurod*. Some regions, also characterized by distinct proliferation zones, show complete absence of *neurog1* expression (subpallium, ventral thalamus, preoptic region, hypothalamus).

Description of levels on facing page

Sensory organs: Strong *neurog1* signal in retinal peripheral edge. Clusters of *neurog1*-positive cells are present in the olfactory epithelium. In both sensory organs, *neurog1* expression domains appear to overlap with proliferating cells.

CNS: In the telencephalon, the pallial domain lies close to, but not directly at the ventricular surface at all anteroposterior levels. Ventral olfactory bulb domains shift increasingly more lateral as the emerging subpallium becomes more prominent in caudal direction. In the diencephalon, large blobs of expression are present in habenula, dorsal thalamus proper and pretectum (all interpreted as alar plate), as well as in basal plate-derived posterior tuberculum (only dorsal division visible on facing page). There is no *neurog1* expression in epiphysis, subpallium and ventral thalamus.

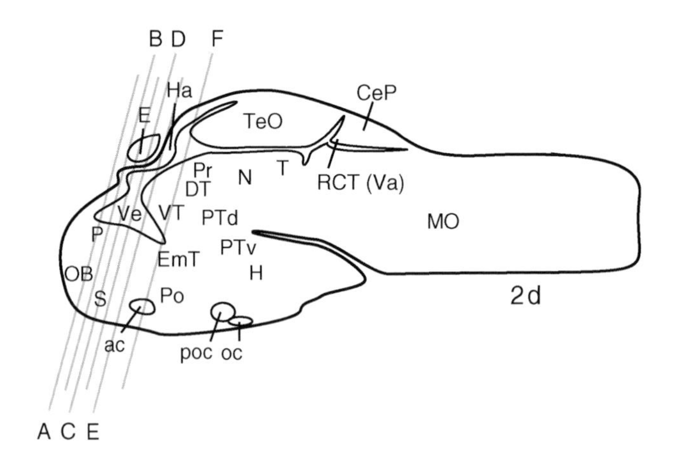

Abbreviations

ac	anterior commissure	OE	olfactory epithelium
CeP	cerebellar plate	P	pallium
DT	dorsal thalamus (thalamus)	pc	posterior commissure
DVe	diencephalic ventricle	Po	preoptic region
E	epiphysis	poc	postoptic commissure
EmT	eminentia thalami	Pr	pretectum
EPi	eye pigment	PTd	dorsal part of posterior tuberculum
H	hypothalamus	PTv	ventral part of posterior tuberculum
Ha	habenula	RCT	rostral cerebellar thickening (valvula)
hac	habenular commissure	S	subpallium
lfb	lateral forebrain bundle	T	midbrain tegmentum
MO	medulla oblongata	TeO	tectum opticum
N	region of the nucleus of medial longitudinal fascicle	TVe	telencephalic ventricle
		Va	valvula cerebelli
OB	olfactory bulb	Ve	brain ventricle
oc	optic chiasma	VT	ventral thalamus (prethalamus)

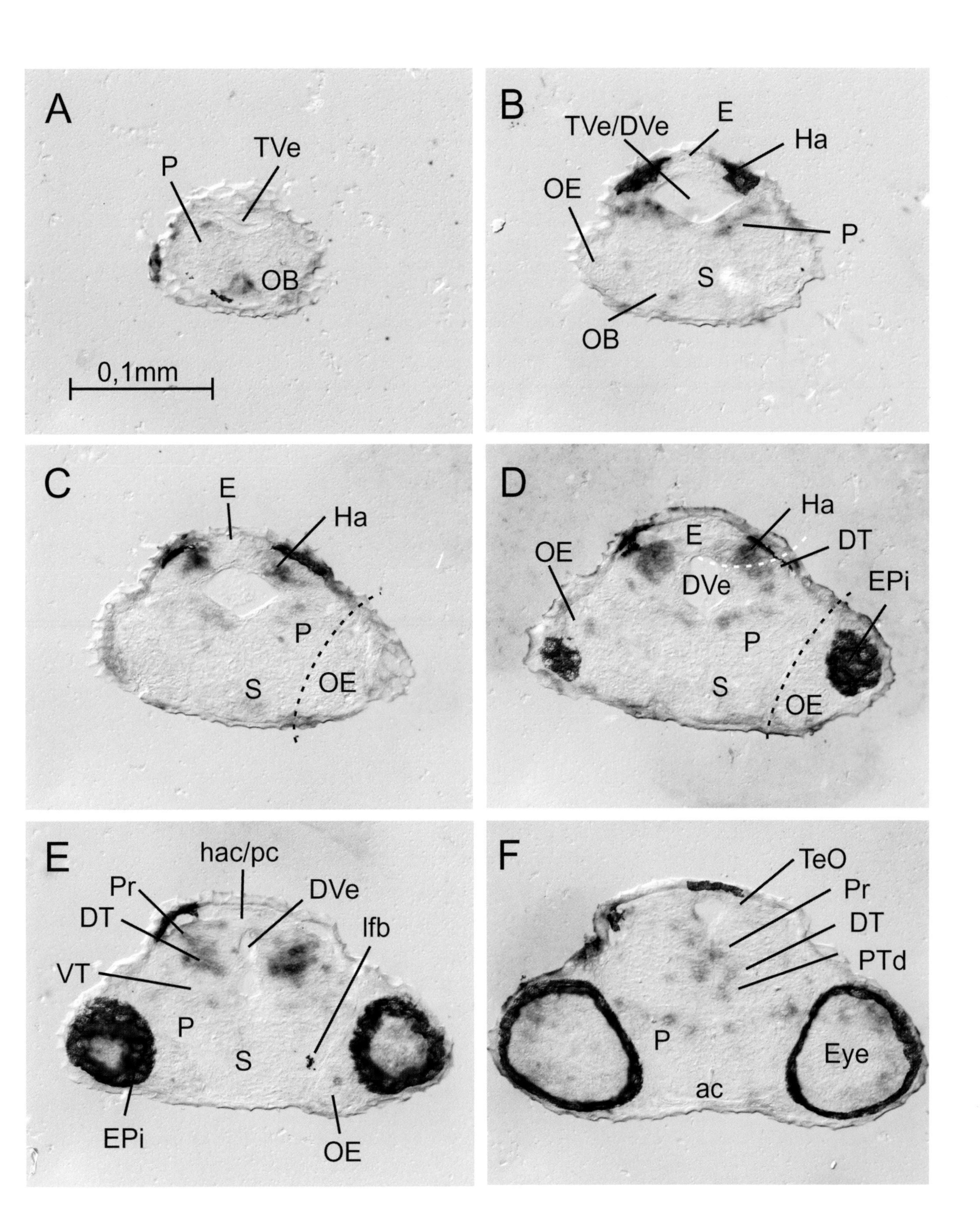
A
P
TVe
OB
0,1mm
B
TVe/DVe
E
Ha
OE
P
S
OB
C
E
Ha
P
S
OE
D
Ha
E
DT
OE
DVe
EPi
P
S
OE
E
hac/pc
Pr
DVe
DT
lfb
VT
P
S
EPi
OE
F
TeO
Pr
DT
PTd
P
Eye
ac

neurog1 (continued, 2nd plate)

Gene Expression Domains in the 2-Day Zebrafish Brain

Description of levels on facing page

Sensory organs: Strong *neurog1* signal in retinal peripheral edge. Retinal expression domains appear to overlap with proliferating cells.

CNS: At these levels, the small domain of the eminentia thalami appears. The subpallium (*neurog1* negative) is caudally replaced by the preoptic region (*neurog1* negative), separated from it by the anterior commissure. A second landmark, the postoptic commissure, separates the preoptic region caudally from the anterior part of the emerging hypothalamus (*neurog1* negative, interpreted as basal plate of the anterior forebrain/secondary prosencephalon). In the diencephalon (posterior forebrain), large blobs of *neurog1* expression are present in dorsal thalamus proper and pretectum (interpreted as alar plate), as well as in posterior tuberculum (dorsal and ventral divisions visible on facing page; interpreted as basal plate). A distinct *neurog1* domain of the midbrain tegmentum is seen, as well as medial, basal (note that *neurog1* cells designated as b may partly lie ventral to tectal ventricle and, thus, belong to medulla oblongata) and lateral domains of the optic tectum. There is no *neurog1* expression in subpallium, preoptic region and hypothalamus.

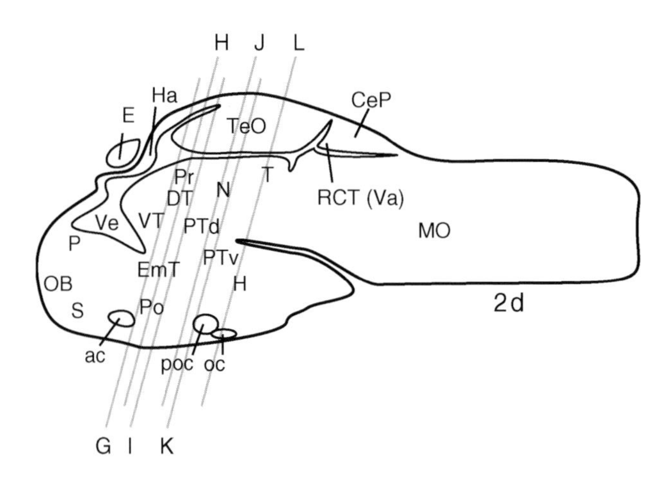

Abbreviations

ac	anterior commissure	oc	optic chiasma
b	basal optic tectum	P	pallium
CeP	cerebellar plate	Po	preoptic region
DT	dorsal thalamus (thalamus)	poc	postoptic commissure
E	epiphysis	Pr	pretectum
EmT	eminentia thalami	PTd	dorsal part of posterior tuberculum
H	hypothalamus	PTv	ventral part of posterior tuberculum
Ha	habenula	RCT	rostral cerebellar thickening (valvula)
Hi	intermediate hypothalamus	S	subpallium
Hr	rostral hypothalamus	T	midbrain tegmentum
l	lateral optic tectum	TeO	tectum opticum
lfb	lateral forebrain bundle	Va	valvula
m	medial optic tectum	Ve	brain ventricle
MO	medulla oblongata	VT	ventral thalamus (prethalamus)
N	region of the nucleus of the medial longitudinal fascicle		
OB	olfactory bulb		

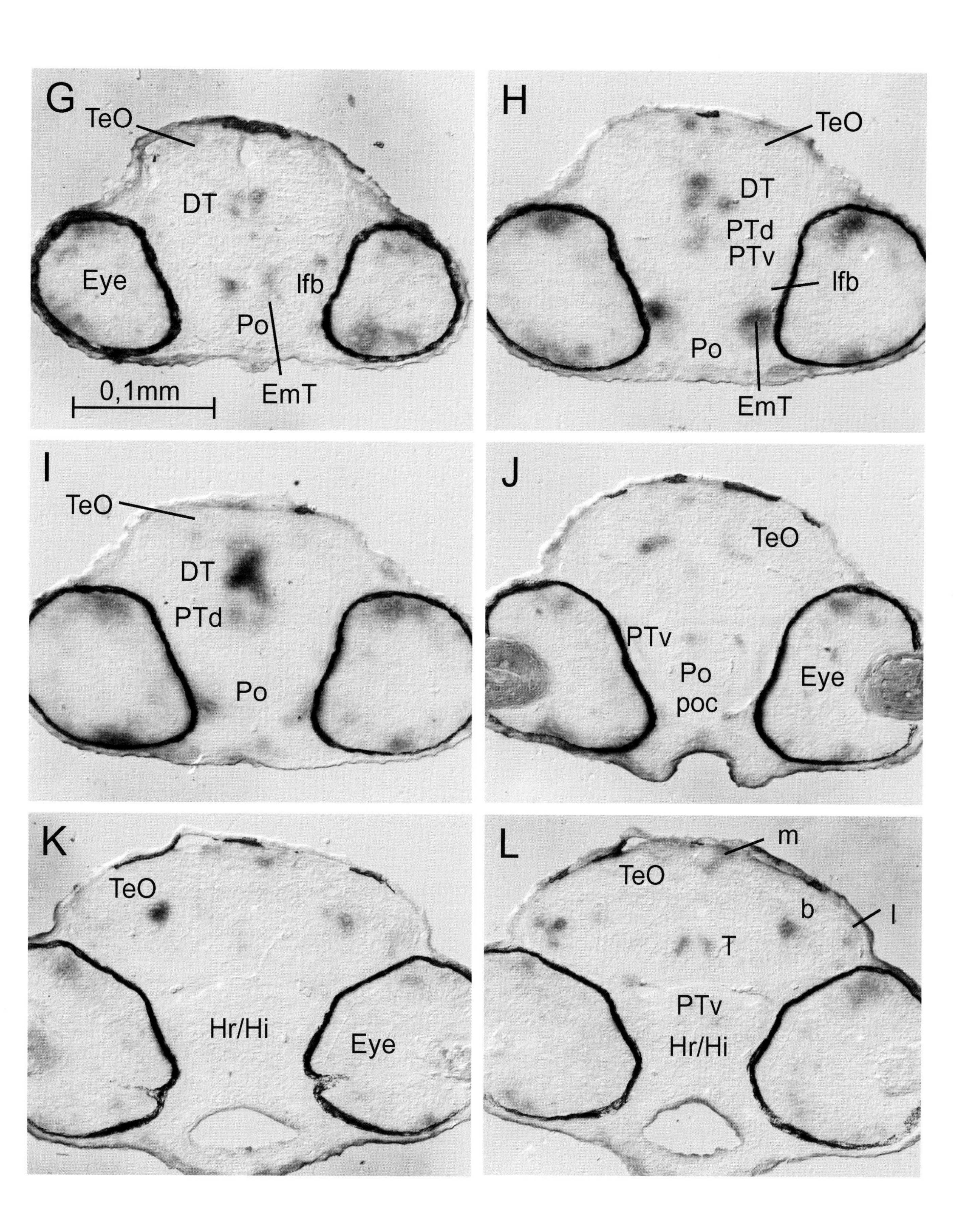
G
TeO
DT
Eye
lfb
Po
EmT
0,1mm
H
TeO
DT
PTd
PTv
lfb
Po
EmT
I
TeO
DT
PTd
Po
J
TeO
PTv
Po
poc
Eye
K
TeO
Hr/Hi
Eye
L
m
TeO
b
l
T
PTv
Hr/Hi

neurog1 (continued, 3rd plate)

Gene Expression Domains in the 2-Day Zebrafish Brain

Description of levels on facing page

Sensory organs/PNS: Strong *neurog1* signal in retinal peripheral edge. Retinal expression domains appear to overlap with proliferating cells.

CNS: Dual composition of inferior lobe is apparent: the ventral, hypothalamic part (interpreted as basal plate of anterior forebrain/secondary prosencephalon) remains *neurog1* negative, whereas the dorsal portion is *neurog1* positive (interpreted as basal plate of ventral thalamic prosomere/P3). Hypophysis shows no *neurog1* signal. Distinct *neurog1* domain of midbrain tegmentum fades out here. Caudally, it is replaced by an extensive *neurog1* domain of the (medullary) midbrain–hindbrain boundary. Medial, basal (note that *neurog1* cells designated as b may partly lie ventral to tectal ventricle and, thus, belong to medulla oblongata) and lateral *neurog1* domains of optic tectum continue at these levels. Most caudal midbrain roof exhibits additionally large *neurog1* domain in torus semicircularis. At this stage, both the rostral part of cerebellum (rostral cerebellar thickening/future valvula) and the more caudal cerebellar plate are *neurog1* negative. Anterior medulla oblongata is strongly *neurog1* positive along the ventricular lining towards the rhombic lip (rim of dorsal opening of medulla oblongata, the rhombic fossa). A most anterior homogeneous medullary *neurog1*-positive cluster is replaced more caudally by various characteristic stripes of *neurog1*-expressing cells emanating ventrally. There is no *neurog1* expression in hypothalamus (permanently).

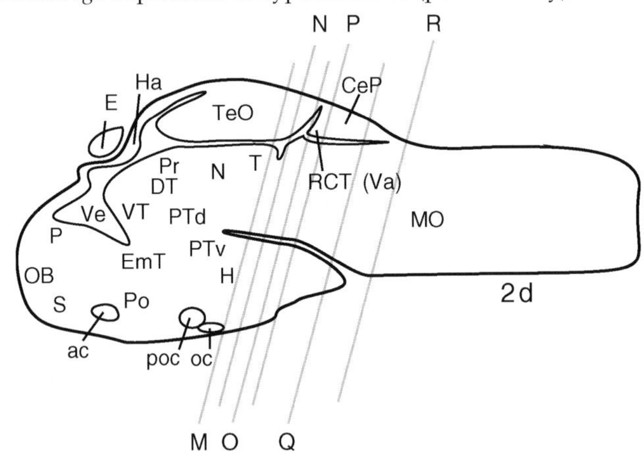

Abbreviations

ac	anterior commissure	OB	olfactory bulb
ALLG	anterior lateral line ganglion	oc	optic chiasma
b	basal optic tectum	P	pallium
Ch	chorda dorsalis	Po	preoptic region
CeP	cerebellar plate	poc	postoptic commissure
DT	dorsal thalamus (thalamus)	Pr	pretectum
E	epiphysis	PTd	dorsal part of posterior tuberculum
EmT	eminentia thalami		
EPi	eye pigment	PTv	ventral part of posterior tuberculum
H	hypothalamus		
Ha	habenula	RCT	rostral cerebellar thickening (valvula)
Hc	caudal hypothalamus		
Hi	intermediate hypothalamus	S	subpallium
Hr	rostral hypothalamus	T	midbrain tegmentum
l	lateral optic tectum	TeO	tectum opticum
LHP	lateral hinge-point between cerebellum/medulla oblongata	TeVe	tectal ventricle
		TG	trigeminal ganglion
m	medial optic tectum	TS	torus semicircularis
MHB	midbrain–hindbrain boundary	Va	valvula cerebelli
MO	medulla oblongata	Ve	brain ventricle
N	region of the nucleus of the medial longitudinal fascicle	VT	ventral thalamus (prethalamus)

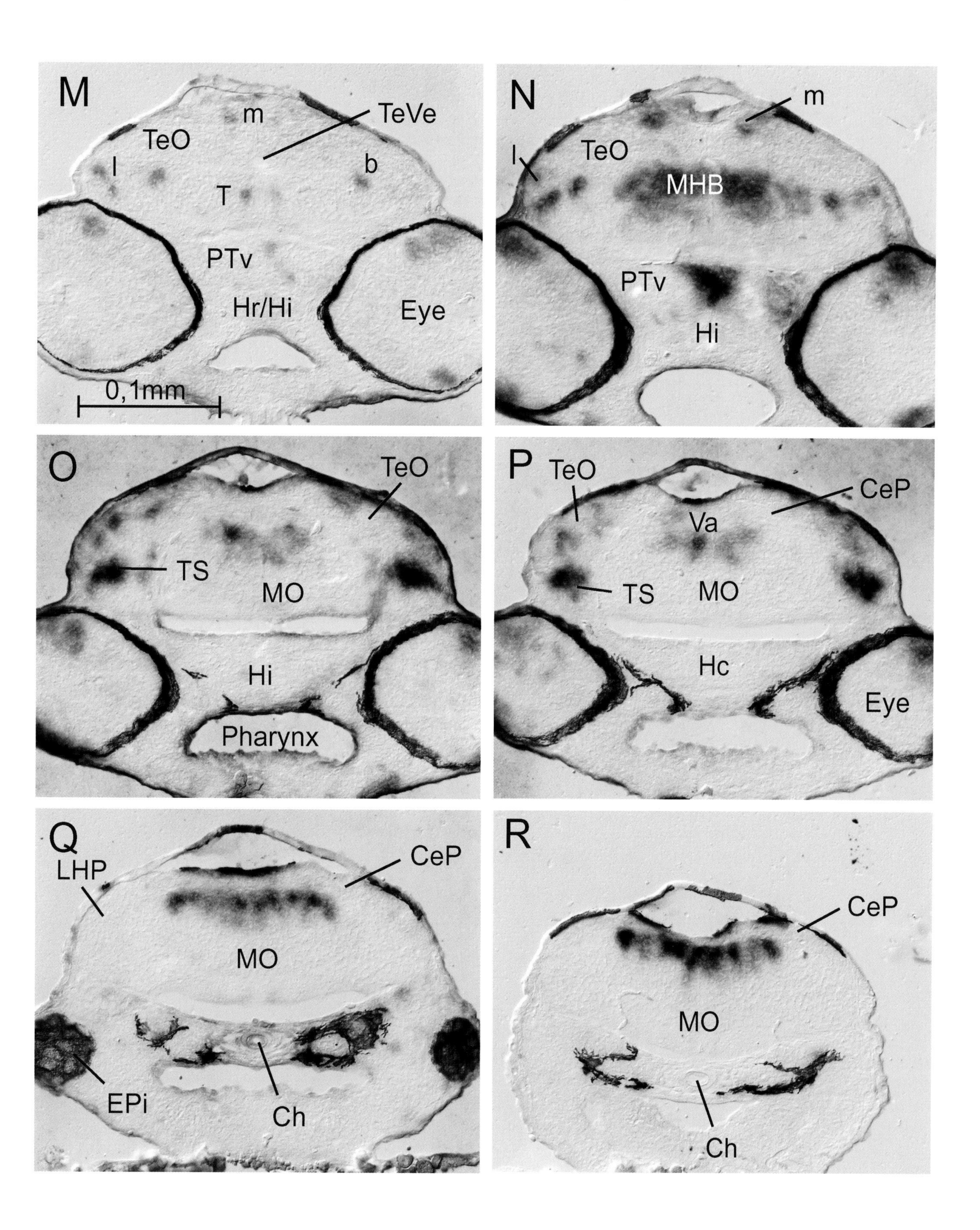

M
TeO
m
TeVe
l
b
T
PTv
Hr/Hi
Eye
0,1mm
N
m
TeO
l
MHB
PTv
Hi
O
TeO
TS
MO
Hi
Pharynx
P
TeO
CeP
Va
TS
MO
Hc
Eye
Q
LHP
CeP
MO
EPi
Ch
R
CeP
MO
Ch

neurog1 (continued, 4th plate)

Gene Expression Domains in the 2-Day Zebrafish Brain

Description of levels on facing page

Sensory organs/PNS: Patchy *neurog1* signal in various cranial nerve ganglia of peripheral nervous system.

CNS: At intermediate medullary levels (characterized by otic capsule), the rhombic lip (rim of dorsal opening of medulla oblongata, the rhombic fossa) is strongly *neurog1* positive. Approaching the obex (boundary to spinal cord), the rhombic lip fuses medially, closing the rhombic fossa. Up to this most caudal level, the medulla oblongata remains *neurog1* positive along the (alar plate) ventricular lining between the bilateral rhombic lips.

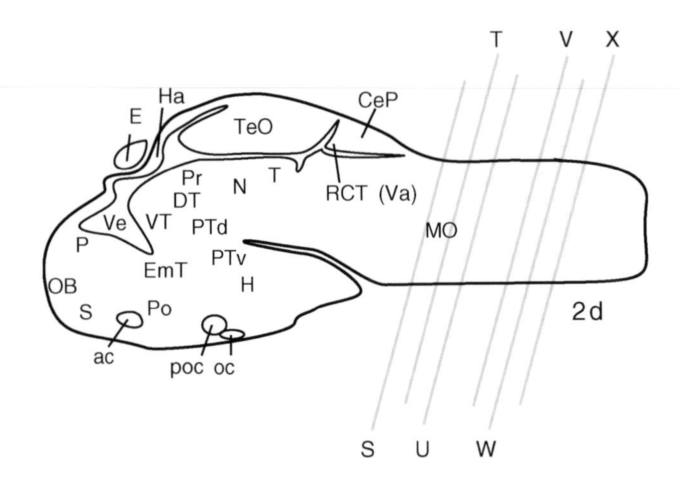

Abbreviations

ac	anterior commissure	Po	preoptic region
CeP	cerebellar plate	poc	postoptic commissure
Ch	chorda dorsalis	Pr	pretectum
DT	dorsal thalamus (thalamus)	PTd	dorsal part of posterior tuberculum
E	epiphysis	PTv	ventral part of posterior tuberculum
EmT	eminentia thalami	RCT	rostral cerebellar thickening (valvula)
FG	facial ganglion	RL	rhombic lip
H	hypothalamus	S	subpallium
Ha	habenula	T	midbrain tegmentum
MO	medulla oblongata	TeO	tectum opticum
N	region of the nucleus of the medial longitudinal fascicle	Va	valvula cerebelli
OB	olfactory bulb	Ve	brain ventricle
OC	otic capsule	VG	vagal ganglion
oc	optic chiasma	VT	ventral thalamus (prethalamus)
OG	octaval ganglion		
P	pallium		

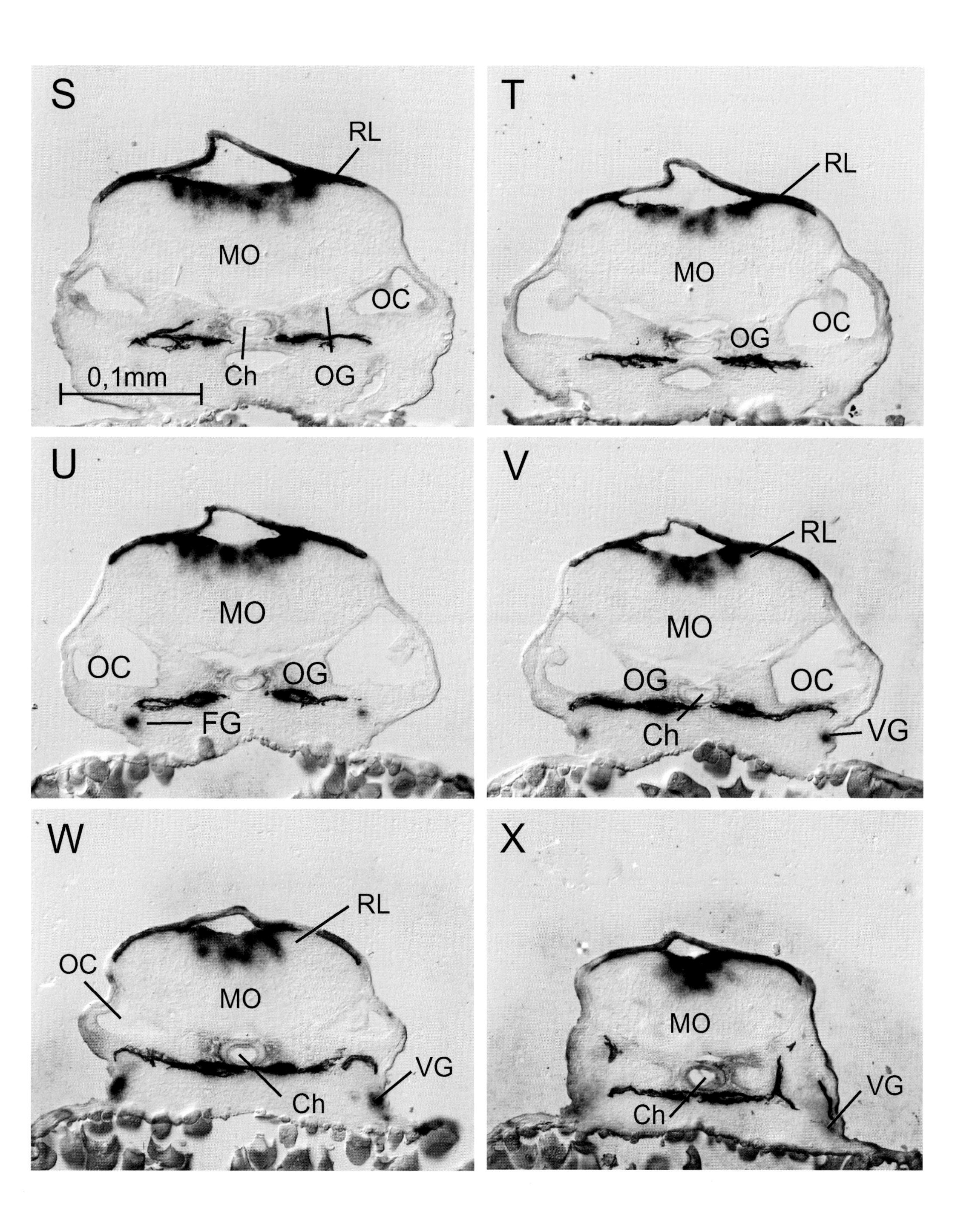
S
RL
MO
OC
Ch
OG
0,1mm
T
RL
MO
OC
OG
U
MO
OC
OG
FG
V
RL
MO
OG
OC
Ch
VG
W
RL
OC
MO
VG
Ch
X
MO
VG
Ch

neurod (ZFIN ID: ZDB-GENE-990415-172; previous names: Beta2/NeuroD; nrd; neuro-D)
Description: Proneural gene, codes for basic Helix–Loop–Helix transcription factor, expressed in early determined neuronal and sensory cells.

Gene Expression Domains in the 2-Day Zebrafish Brain

General appearance: Consistent with a role in early neuronal determination, *neurod* expression domains lie one to several cell rows away from the brain ventricle, flanked by both *neurod*-negative proliferation zones towards the ventricle and more peripheral differentiating cell masses towards the pia. Expression patterns parallel those revealed with proliferation markers (e.g., prosomeric organization in posterior forebrain, i.e., pretectum, dorsal thalamus, ventral thalamus and rhombomeric organization in medulla oblongata), including PNS. Some regions, also characterized by distinct proliferation zones, show complete absence of *neurod* expression (subpallium, ventral thalamus, preoptic region, hypothalamus).

Description of levels on facing page

Sensory organs: Strong *neurod* signal in outer and inner nuclear layers of central retina, but not in ganglionic layer and retinal peripheral edge. Clusters of *neurod*-positive cells are present in the olfactory epithelium.

CNS: In the telencephalon, large pallial domain lies somewhat remote from the ventricular surface at all anteroposterior levels. Ventral olfactory bulb domains shift increasingly more lateral as the emerging subpallium becomes more prominent in caudal direction. In the diencephalon, large blobs of expression are present in epiphysis, habenula, dorsal thalamus proper and pretectum (all interpreted as alar plate), as well as in basal plate-derived posterior tuberculum (only dorsal division visible on facing page). The anterior optic tectum contains a medial *neurod* domain; the large domain below the anterior tectal ventricle may belong to the griseum tectale. There is no *neurod* expression in subpallium and ventral thalamus.

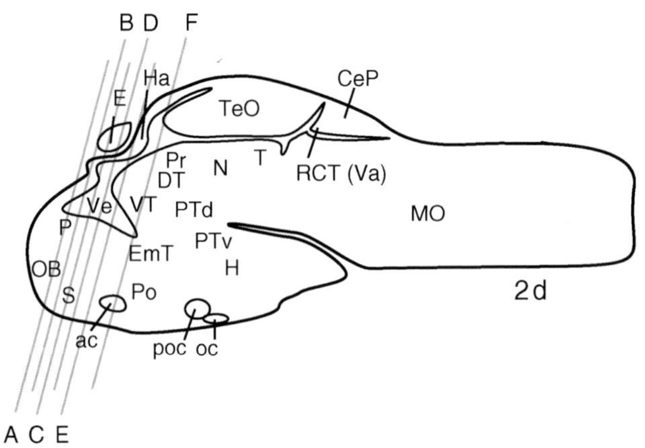

Abbreviations

ac	anterior commissure	OE	olfactory epithelium
CeP	cerebellar plate	P	pallium
DT	dorsal thalamus (thalamus)	pc	posterior commissure
DVe	diencephalic ventricle	Po	preoptic region
E	epiphysis	poc	postoptic commissure
EmT	eminentia thalami	Pr	pretectum
EPi	eye pigment	PTd	dorsal part of posterior tuberculum
GT	griseum tectale	PTv	ventral part of posterior tuberculum
H	hypothalamus	RCT	rostral cerebellar thickening (valvula)
Ha	habenula	S	subpallium
hac	habenular commissure	T	midbrain tegmentum
MO	medulla oblongata	TeO	tectum opticum
N	region of the nucleus of medial longitudinal fascicle	TVe	telencephalic ventricle
		Va	valvula cerebelli
OB	olfactory bulb	Ve	brain ventricle
oc	optic chiasma	VT	ventral thalamus (prethalamus)

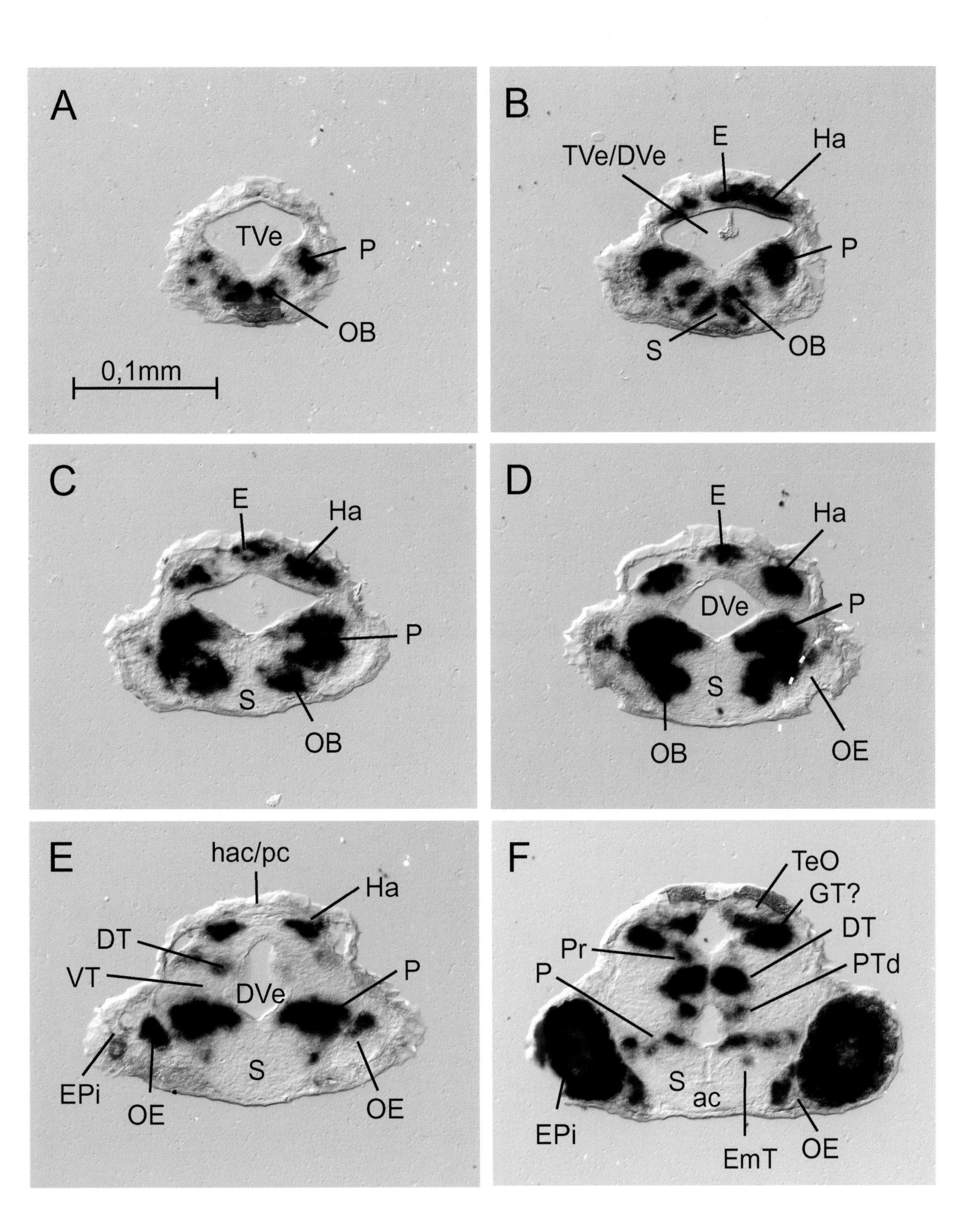
A
TVe
P
OB
0,1mm
B
E
Ha
TVe/DVe
P
S
OB
C
E
Ha
P
S
OB
D
E
Ha
DVe
P
S
OB
OE
E
hac/pc
Ha
DT
VT
DVe
P
S
EPi
OE
OE
F
TeO
GT?
DT
Pr
PTd
P
S
ac
EPi
EmT
OE

neurod (continued, 2nd plate)

Gene Expression Domains in the 2-Day Zebrafish Brain

Description of levels on facing page

Sensory organs: Strong *neurod* signal in outer and inner nuclear layers of central retina, but not in ganglionic layer and retinal peripheral edge.

CNS: At these levels, the pallial *neurod* domain fades out. In its wake, the small domain of the eminentia thalami appears. The subpallium (*neurod* negative) is caudally replaced by the preoptic region (*neurod* negative), separated from it by the anterior commissure. A second landmark, the postoptic commissure, separates the preoptic region caudally from the anterior part of the emerging hypothalamus (*neurod* negative, interpreted as basal plate of the anterior forebrain/secondary prosencephalon). In the diencephalon (posterior forebrain), large blobs of *neurod* expression are present in dorsal thalamus proper and pretectum (interpreted as alar plate), as well as in posterior tuberculum (dorsal and ventral divisions visible on facing page; interpreted as basal plate). Caudal to the posterior tuberculum, a small *neurod* blob appears in basal synencephalon (basal plate of pretectal prosomere/P1), the region of the nucleus of the medial longitudinal fascicle. A distinct *neurod* domain of the midbrain tegmentum is seen, as well as the large medial, basal and lateral domains of the optic tectum. There is no *neurod* expression in subpallium, preoptic region and hypothalamus.

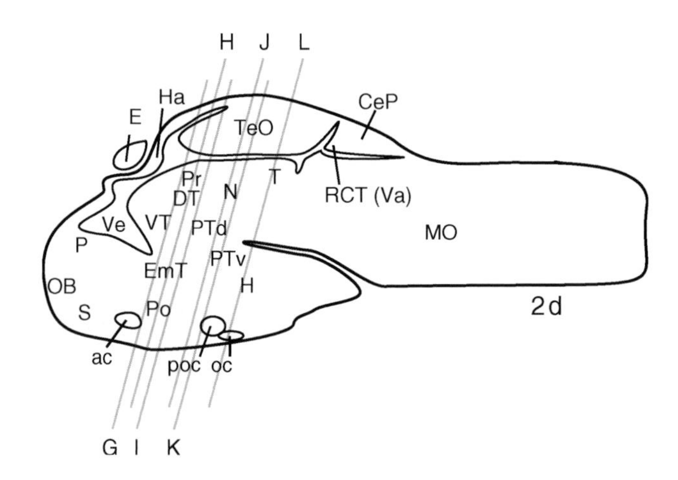

Abbreviations

ac	anterior commissure	P	pallium
CeP	cerebellar plate	Po	preoptic region
DT	dorsal thalamus (thalamus)	poc	postoptic commissure
E	epiphysis	Pr	pretectum
EmT	eminentia thalami	PTd	dorsal part of posterior tuberculum
H	hypothalamus	PTv	ventral part of posterior tuberculum
Ha	habenula	RCT	rostral cerebellar thickening (valvula)
Hi	intermediate hypothalamus	S	subpallium
Hr	rostral hypothalamus	T	midbrain tegmentum
lfb	lateral forebrain bundle	TeO	tectum opticum
MO	medulla oblongata	Va	valvula
N	region of the nucleus of the medial longitudinal fascicle	Ve	brain ventricle
OB	olfactory bulb	VT	ventral thalamus (prethalamus)
oc	optic chiasma		

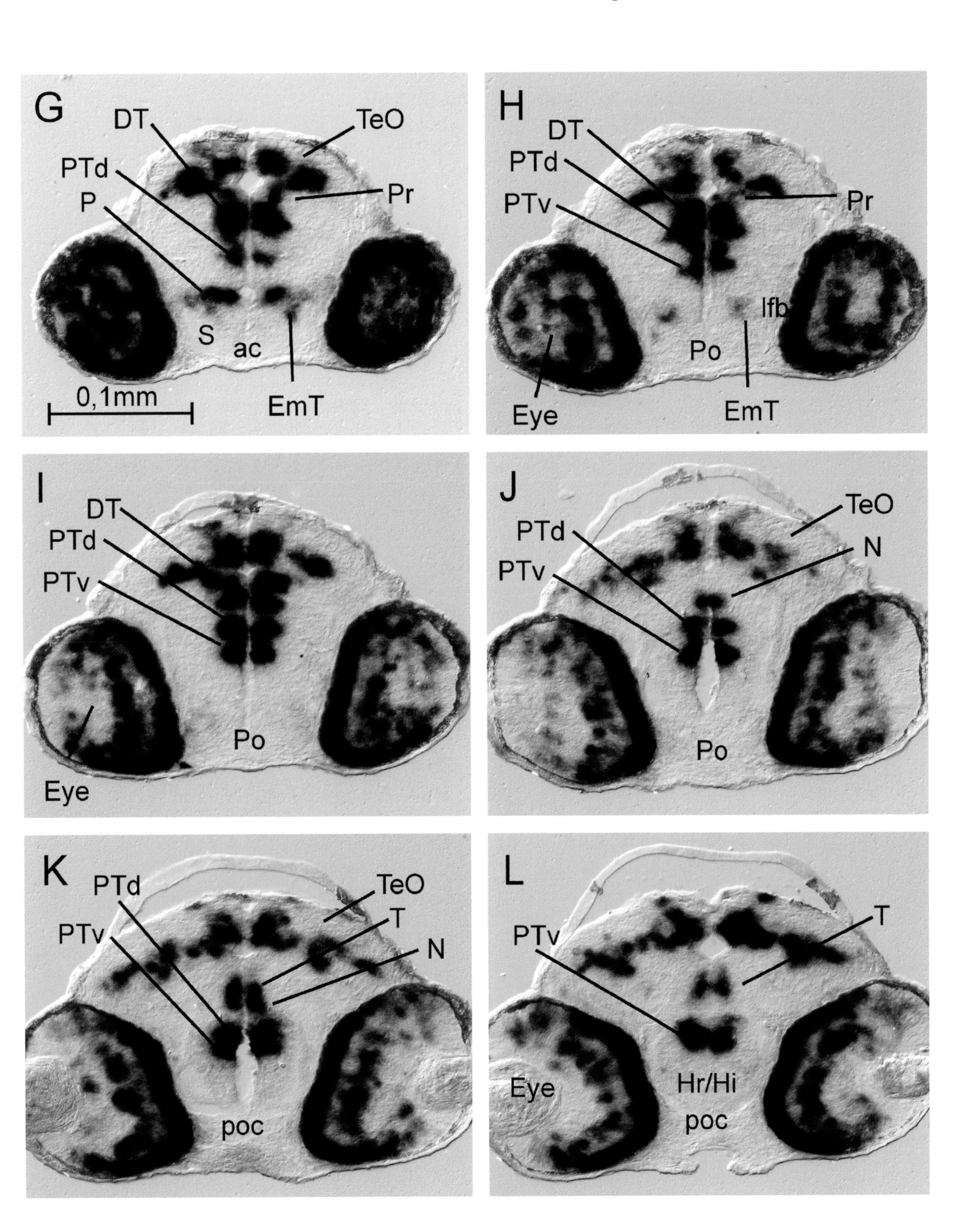
G
DT
TeO
PTd
P
Pr
S
ac
0,1mm
EmT
H
DT
PTd
PTv
Pr
lfb
Po
Eye
EmT
I
DT
PTd
PTv
Po
Eye
J
TeO
PTd
N
PTv
Po
K
PTd
TeO
PTv
T
N
poc
L
T
PTv
Eye
Hr/Hi
poc

neurod (continued, 3rd plate)

Gene Expression Domains in the 2-Day Zebrafish Brain

Description of levels on facing page

Sensory organs/PNS: Strong *neurod* signal in outer and inner nuclear layers of central retina, but not in ganglionic layer and retinal peripheral edge. Expression of *neurod* is also present in cranial nerve ganglia of peripheral nervous system.

CNS: Dual composition of inferior lobe is apparent: the ventral, hypothalamic part (interpreted as basal plate of anterior forebrain/secondary prosencephalon) remains *neurod* negative, whereas the dorsal portion is *neurod* positive (interpreted as basal plate of ventral thalamic prosomere/P3). Hypophysis shows strong *neurod* signal. Distinct *neurod* domain of midbrain tegmentum fades out towards caudally adjacent, strongly proliferative midbrain–hindbrain boundary (*neurod* negative). Large medial, basal and lateral *neurod* domains of optic tectum continue at these levels. Most caudal midbrain roof exhibits additionally large *neurod* domain in torus semicircularis. Rostral part of cerebellum (rostral cerebellar thickening/future valvula) is *neurod* negative, but more caudal cerebellar plate is strongly *neurod* positive. Anterior medulla oblongata is strongly *neurod* positive, especially along rhombic lip (rim of dorsal opening of medulla oblongata, the rhombic fossa). Many *neurod*-positive cells apparently migrate ventrally from rhombic lip, especially at the lateral hinge-point region of cerebellar plate/rhombic lip. In the medial (alar plate) medulla oblongata, a most anterior homogeneous *neurod*-positive cluster is replaced more caudally by various characteristic stripes of *neurod*-expressing cells emanating ventrally. There is no *neurod* expression in hypothalamus (permanently) and in midbrain–hindbrain boundary, including valvula cerebelli (temporary).

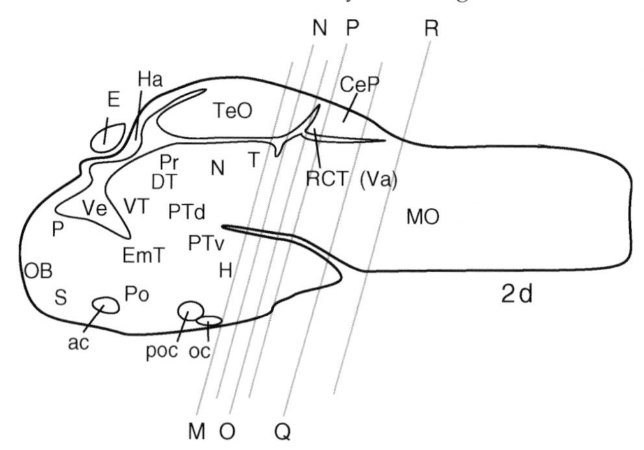

Abbreviations

ac	anterior commissure	OB	olfactory bulb
ALLG	anterior lateral line ganglion	oc	optic chiasma
CeP	cerebellar plate	P	pallium
Ch	chorda dorsalis	Po	preoptic region
DT	dorsal thalamus (thalamus)	poc	postoptic commissure
E	epiphysis	Pr	pretectum
EmT	eminentia thalami	PTd	dorsal part of posterior tuberculum
H	hypothalamus	PTv	ventral part of posterior tuberculum
Ha	habenula	RCT	rostral cerebellar thickening (valvula)
Hc	caudal hypothalamus		
Hi	intermediate hypothalamus	RL	rhombic lip
Hr	rostral hypothalamus	S	subpallium
Hy	hypophysis (pituitary)	T	midbrain tegmentum
LHP	lateral hinge-point between cerebellum/medulla oblongata	TeO	tectum opticum
		TG	trigeminal ganglion
MHB	midbrain–hindbrain boundary	TS	torus semicircularis
MO	medulla oblongata	Va	valvula cerebelli
N	region of the nucleus of the medial longitudinal fascicle	Ve	brain ventricle
		VT	ventral thalamus (prethalamus)

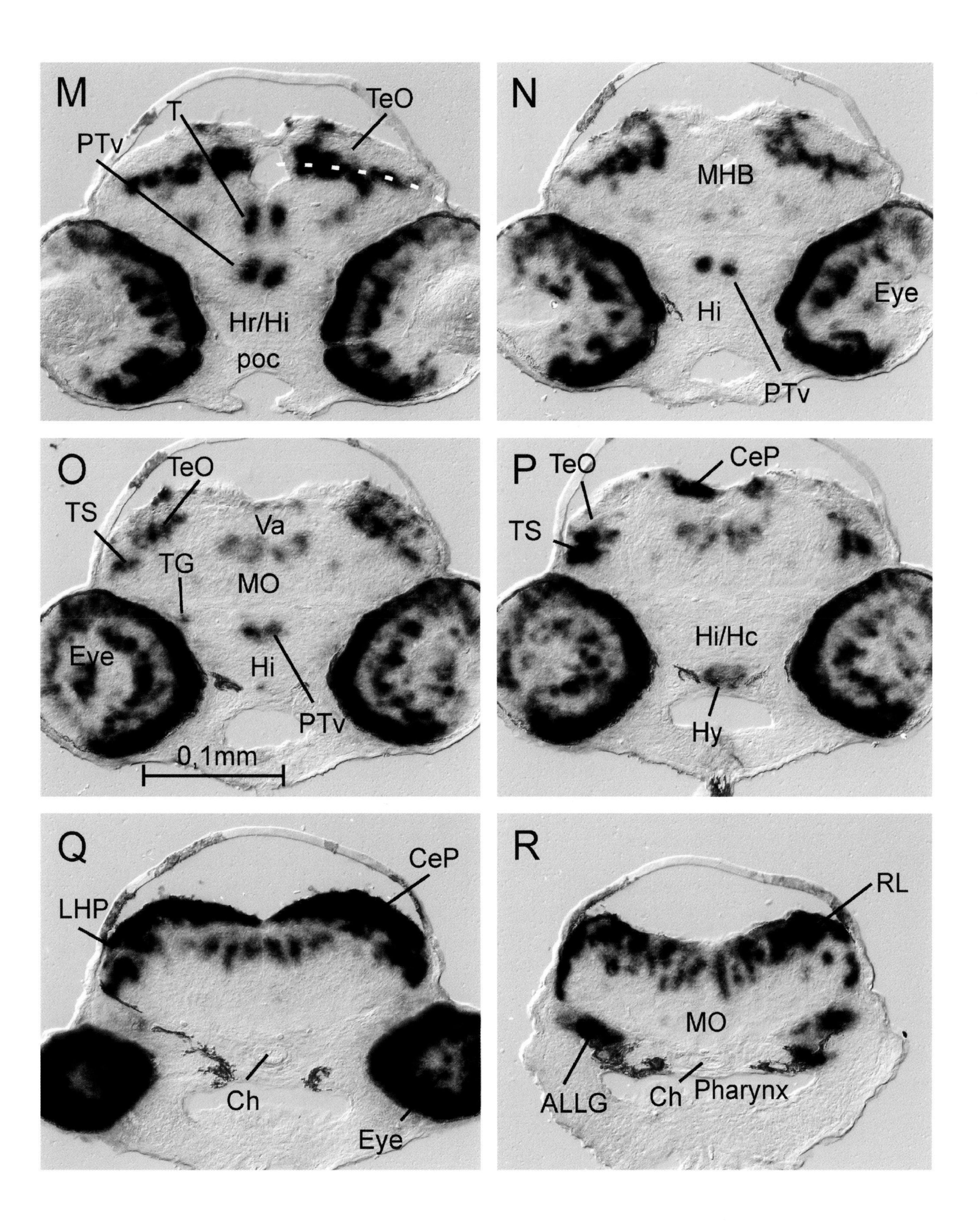
M
T
TeO
PTv
Hr/Hi
poc
N
MHB
Hi
Eye
PTv
O
TeO
TS
Va
TG
MO
Eye
Hi
PTv
0,1mm
P
TeO
CeP
TS
Hi/Hc
Hy
Q
CeP
LHP
Ch
Eye
R
RL
MO
ALLG
Ch
Pharynx

neurod (continued, 4th plate)

Gene Expression Domains in the 2-Day Zebrafish Brain

Description of levels on facing page

Sensory organs/PNS: Patchy *neurod* signal in cellular lining of otic capsule (future stato-acoustic ganglion cells) and expression of *neurod* in various cranial nerve ganglia of peripheral nervous system.

CNS: At intermediate medullary levels (characterized by otic capsule), the rhombic lip (rim of dorsal opening of medulla oblongata, the rhombic fossa) remains strongly *neurod* positive, with many expressing cells emanating from it ventrally at the lateral periphery of the rhombencephalon. Approaching the obex (boundary to spinal cord), the rhombic lip fuses medially, closing the rhombic fossa. Up to this most caudal level, the medulla oblongata remains *neurod* positive in the (alar plate) midline (as is the more anterior rhombencephalon). At intermediate to caudal medullary levels, various distinct clusters of *neurod*-positive cells emanate ventrally into the rhombencephalic gray matter.

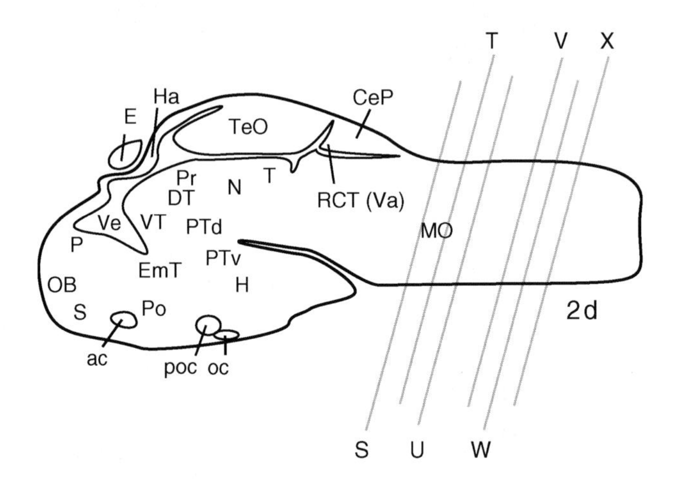

Abbreviations

ac	anterior commissure	PLLG	posterior lateral line ganglion
CeP	cerebellar plate	Po	preoptic region
Ch	chorda dorsalis	poc	postoptic commissure
DT	dorsal thalamus (thalamus)	Pr	pretectum
E	epiphysis	PTd	dorsal part of posterior tuberculum
EmT	eminentia thalami	PTv	ventral part of posterior tuberculum
FG	facial ganglion	RCT	rostral cerebellar thickening (valvula)
GG	glossopharyngeal ganglion	RL	rhombic lip
H	hypothalamus	S	subpallium
Ha	habenula	T	midbrain tegmentum
MO	medulla oblongata	TeO	tectum opticum
N	region of the nucleus of the medial longitudinal fascicle	Va	valvula cerebelli
OB	olfactory bulb	Ve	brain ventricle
OC	otic capsule	VG	vagal ganglion
oc	optic chiasma	VT	ventral thalamus (prethalamus)
OG	octaval ganglion		
P	pallium		

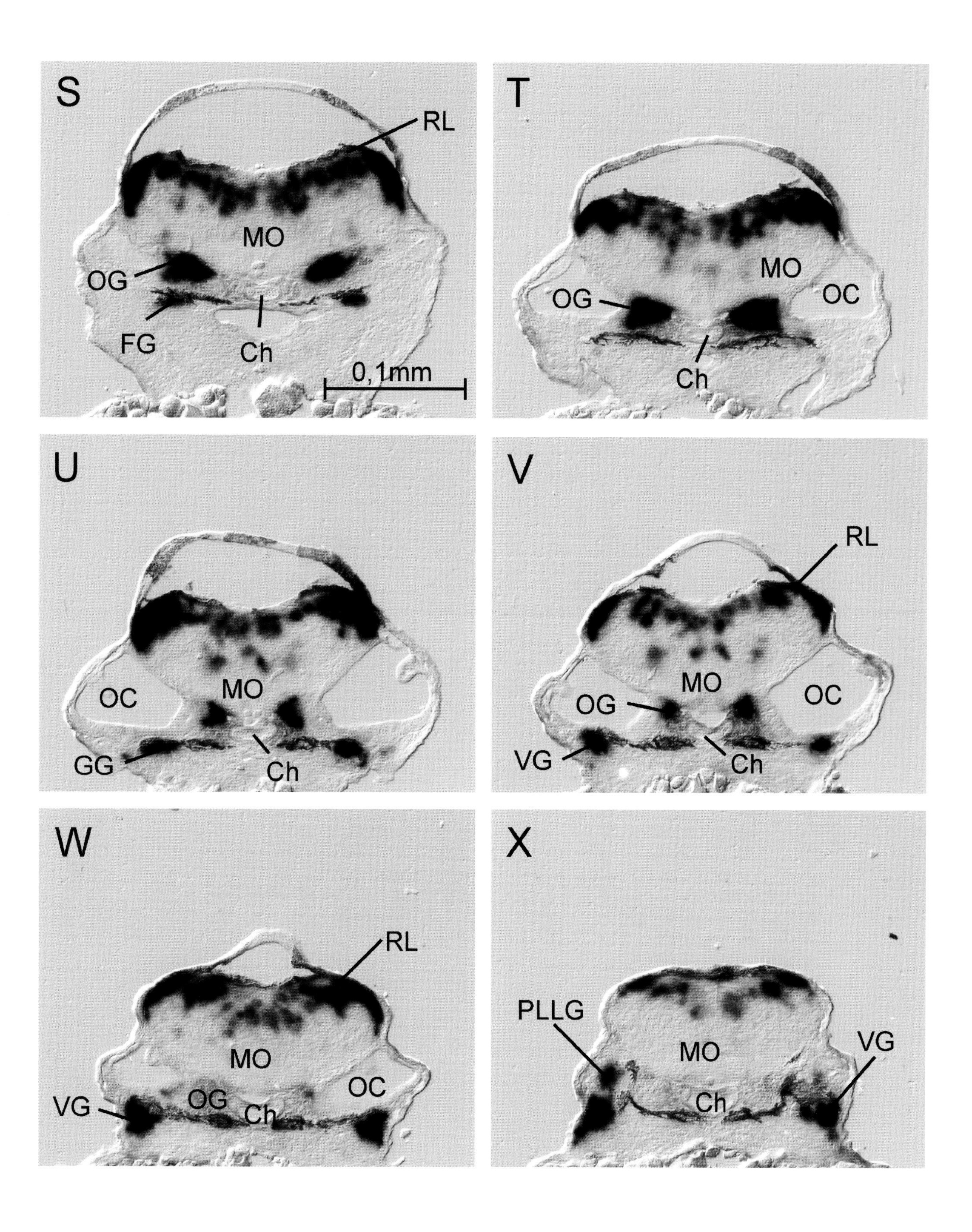
S
RL
MO
OG
FG
Ch
0,1mm
T
MO
OC
OG
Ch
U
OC
MO
GG
Ch
V
RL
MO
OC
OG
VG
Ch
W
RL
MO
OC
VG
OG
Ch
X
PLLG
MO
VG
Ch

Hu-proteins (shown with anti-Hu monoclonal antibody 16A11; source: Monoclonal Antibody Facility, Eugene, OR)
Description: mRNA binding proteins with a gene-regulatory function at the posttranscriptional level (Marusich et al. 1994). Antibody used recognizes peptide sequence present in various Hu-proteins, including those specifically expressed in neuronal cells (i.e., HuD, HuC; Park et al. 2000). Expression in early differentiating neuronal cells.

Protein Expression Domains in the 2-Day Zebrafish Brain

General appearance: Consistent with a role in early neuronal differentiation, Hu expression domains lie away from the ventricle (i.e., lateral to proliferative, *notch1a*-positive cells and early determined neuronal, *neurod*-positive cells), in the most peripheral gray matter of the entire CNS, as well as in the PNS. Regions with ongoing strong proliferation show less Hu-positive cells with weaker staining compared with more mature regions.

Description of levels on facing page

Sensory organs: At these levels, the retina is free of label. Many strongly Hu-positive cells are seen in the olfactory epithelium.

CNS: In the telencephalon, Hu-stained cells are seen in the olfactory bulb, as well as in the most peripheral gray matter of pallium and subpallium. The large pallial Hu domain lies remote from the ventricular surface at all anteroposterior levels. In the diencephalon, Hu-positive cells are seen in epiphysis and a few in habenula and pretectum, while ventral and dorsal thalamus proper remain free of label at these levels (all interpreted as alar plate). Some Hu-positive cells are seen in most anterior, peripheral optic tectum.

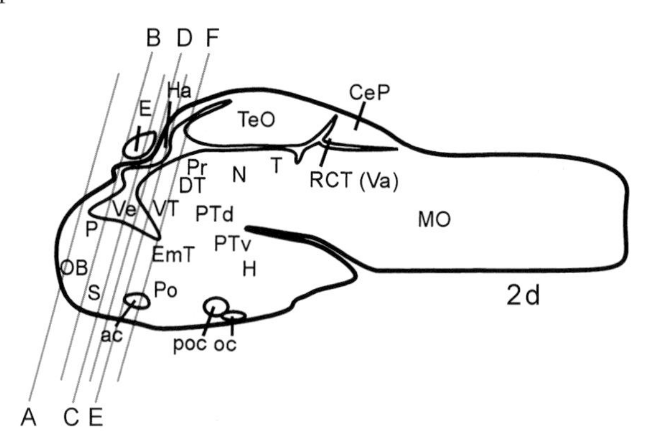

Abbreviations

ac	anterior commissure	pc	posterior commissure
CeP	cerebellar plate	Po	preoptic region
DT	dorsal thalamus (thalamus)	poc	postoptic commissure
DVe	diencephalic ventricle	Pr	pretectum
E	epiphysis	PTd	dorsal part of posterior tuberculum
EmT	eminentia thalami	PTv	ventral part of posterior tuberculum
EPi	eye pigment	RCT	rostral cerebellar thickening (valvula)
H	hypothalamus	S	subpallium
Ha	habenula	T	midbrain tegmentum
hac	habenular commissure	TeO	tectum opticum
MO	medulla oblongata	TVe	telencephalic ventricle
N	region of the nucleus of medial longitudinal fascicle	Va	valvula cerebelli
OB	olfactory bulb	Ve	brain ventricle
oc	optic chiasma	VT	ventral thalamus (prethalamus)
OE	olfactory epithelium		
P	pallium		

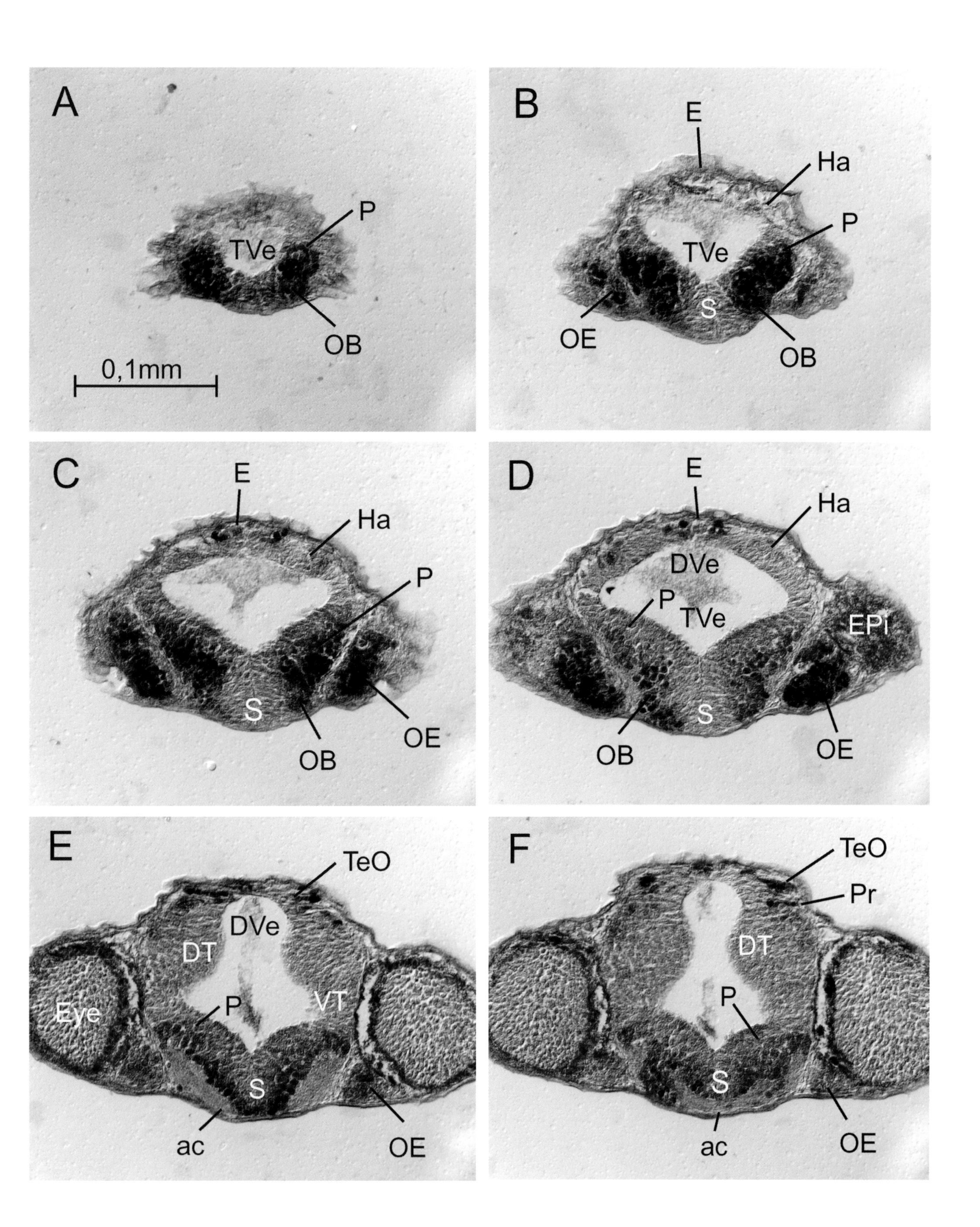
A
TVe
P
OB
0,1mm
B
E
Ha
P
TVe
S
OE
OB
C
E
Ha
P
S
OB
OE
D
E
Ha
DVe
P
TVe
EPi
S
OB
OE
E
TeO
DVe
DT
Eye
P
VT
S
ac
OE
F
TeO
Pr
DT
P
S
ac
OE

Hu-proteins (continued, 2nd plate)

Protein Expression Domains in the 2-Day Zebrafish Brain

Description of levels on facing page

Sensory organs: Hu-protein is present in ganglionic layer of retina.

CNS: Hu-positive cells are seen in peripheral area of eminentia thalami (future entopeduncular complex) and preoptic region, which replaces the subpallium caudal to the anterior commissure. In the diencephalon (posterior forebrain), Hu-stained cells are present at these levels in the dorsal and ventral thalami and pretectum (interpreted as alar plate), as well as in the posterior tuberculum (dorsal and ventral divisions visible on facing page; interpreted as basal plate) and in the region of the basal synencephalon (i.e., nucleus of the medial longitudinal fascicle; basal plate). Note the peripheral Hu-positive cells in the far migrated region M2 (future preglomerular complex). Another landmark, the postoptic commissure, separates the preoptic region caudally from the rostral part of the emerging rostral hypothalamus (interpreted as basal plate of the anterior forebrain/secondary prosencephalon) which also exhibits peripheral Hu-positive cells. Caudal to posterior tuberculum and basal synencephalon, large peripheral Hu-positive cell masses are seen in the midbrain tegmentum. In the optic tectum, few Hu-positive cells are seen, mostly anteriorly; the Hu-positive cells arching below the tectal ventricle may correspond to the griseum tectale (alar plate).

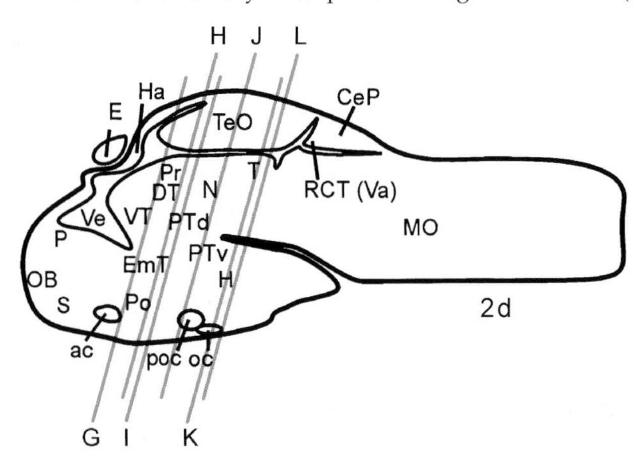

Abbreviations

ac	anterior commissure	OR	optic recess
CeP	cerebellar plate	P	pallium
DT	dorsal thalamus (thalamus)	pc	posterior commissure
E	epiphysis	Po	preoptic region
EmT	eminentia thalami	poc	postoptic commissure
GT	griseum tectale	Pr	pretectum
H	hypothalamus	PTd	dorsal part of posterior tuberculum
Ha	habenula		
Hi	intermediate hypothalamus	PTv	ventral part of posterior tuberculum
Hr	rostral hypothalamus		
lfb	lateral forebrain bundle	RCT	rostral cerebellar thickening (valvula)
M2	migrated posterior tubercular area		
MO	medulla oblongata	S	subpallium
N	region of the nucleus of the medial longitudinal fascicle	T	midbrain tegmentum
		TeO	tectum opticum
OB	olfactory bulb	Va	valvula
oc	optic chiasma	Ve	brain ventricle
on	optic nerve	VT	ventral thalamus (prethalamus)

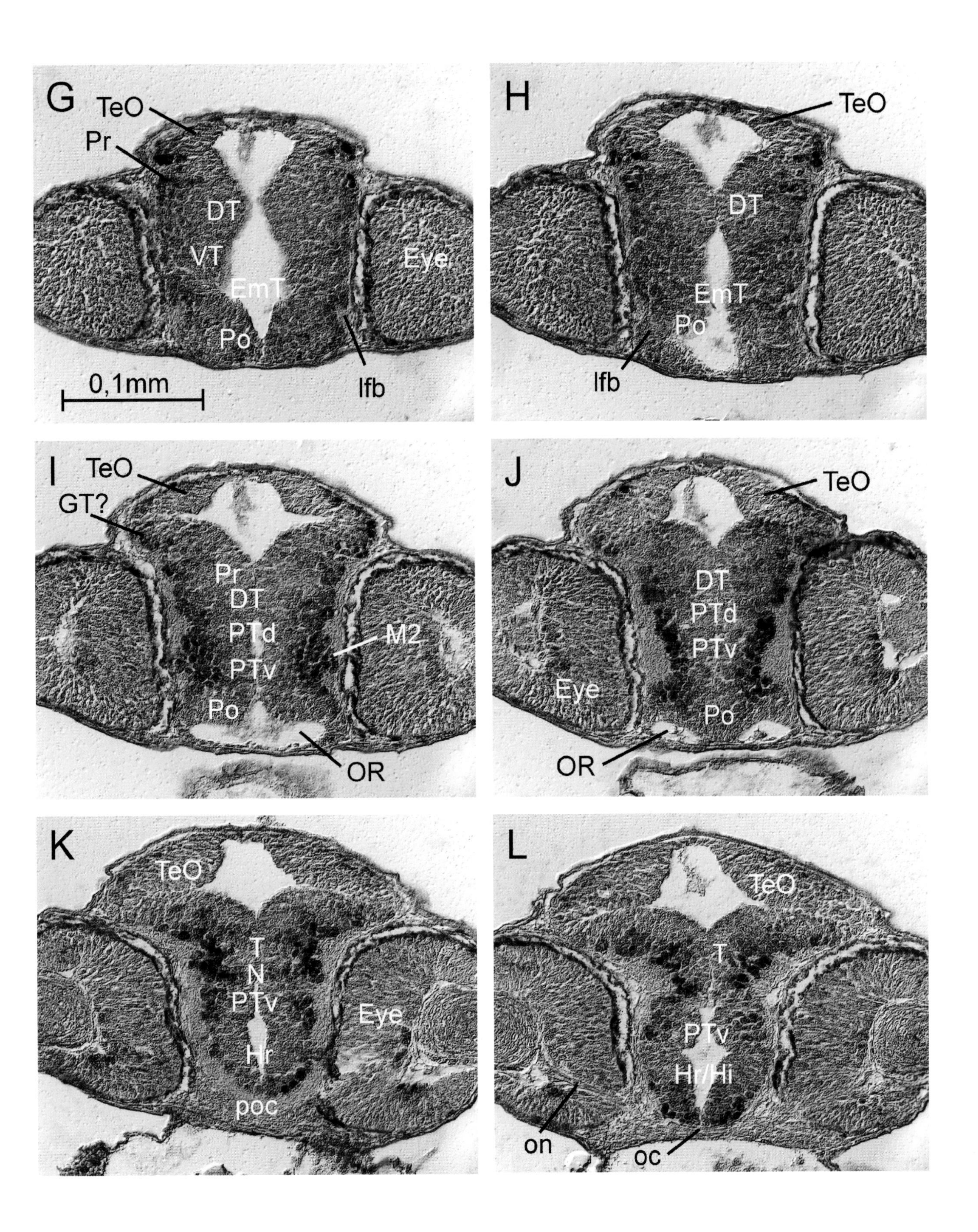
G
TeO
Pr
DT
VT
EmT
Po
Eye
0,1mm
lfb
H
TeO
DT
EmT
Po
lfb
I
TeO
GT?
Pr
DT
PTd
PTv
Po
M2
OR
J
TeO
DT
PTd
PTv
Eye
Po
OR
K
TeO
T
N
PTv
Hr
Eye
poc
L
TeO
T
PTv
Hr/Hi
on
oc

Hu-proteins (continued, 3rd plate)

Protein Expression Domains in the 2-Day Zebrafish Brain

Description of levels on facing page

Sensory organs/PNS: Hu-protein is present in ganglionic layer of retina. Expression of Hu is also present in cranial nerve ganglia of peripheral nervous system.

CNS: Both ventral division of inferior lobe (hypothalamic part; interpreted as basal plate of anterior forebrain/secondary prosencephalon) and dorsal portion (interpreted as basal plate of ventral thalamic prosomere/P3) exhibit Hu-positive cells, except for the posterior inferior lobe and the hypophysis. Large Hu-positive cell masses of midbrain tegmentum thin out towards caudally adjacent, strongly proliferative midbrain–hindbrain boundary. At these levels, the optic tectum, as well as the torus semicircularis and entire cerebellum (valvula and cerebellar plate) remain free of Hu stain. The medulla oblongata is strongly Hu positive in the periphery, but not the rhombic lip (rim of dorsal opening of medulla oblongata, the rhombic fossa) and not the floor plate cells.

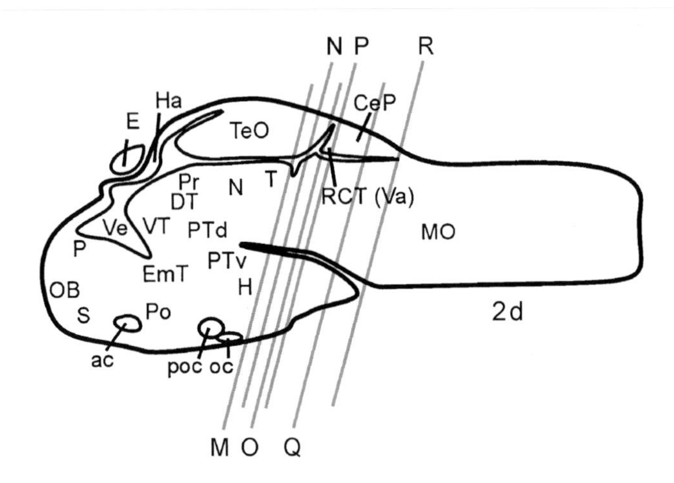

Abbreviations

ac	anterior commissure	OB	olfactory bulb
ALLG	anterior lateral line ganglion	oc	optic chiasma
CeP	cerebellar plate	P	pallium
Ch	chorda dorsalis	Po	preoptic region
DT	dorsal thalamus (thalamus)	poc	postoptic commissure
E	epiphysis	Pr	pretectum
FG	facial ganglion	PTd	dorsal part of posterior tuberculum
FP	floor plate	PTv	ventral part of posterior tuberculum
EmT	eminentia thalami	RCT	rostral cerebellar thickening (valvula)
H	hypothalamus	RL	rhombic lip
Ha	habenula	S	subpallium
Hc	caudal hypothalamus	T	midbrain tegmentum
Hi	intermediate hypothalamus	TeO	tectum opticum
Hr	rostral hypothalamus	TG	trigeminal ganglion
Hy	hypophysis (pituitary)	TS	torus semicircularis
LHP	lateral hinge-point between cerebellum/medulla oblongata	Va	valvula cerebelli
LVe	lateral recess ventricle of H	Ve	brain ventricle
MHB	midbrain–hindbrain boundary	VT	ventral thalamus (prethalamus)
MO	medulla oblongata		
N	region of the nucleus of the medial longitudinal fascicle		

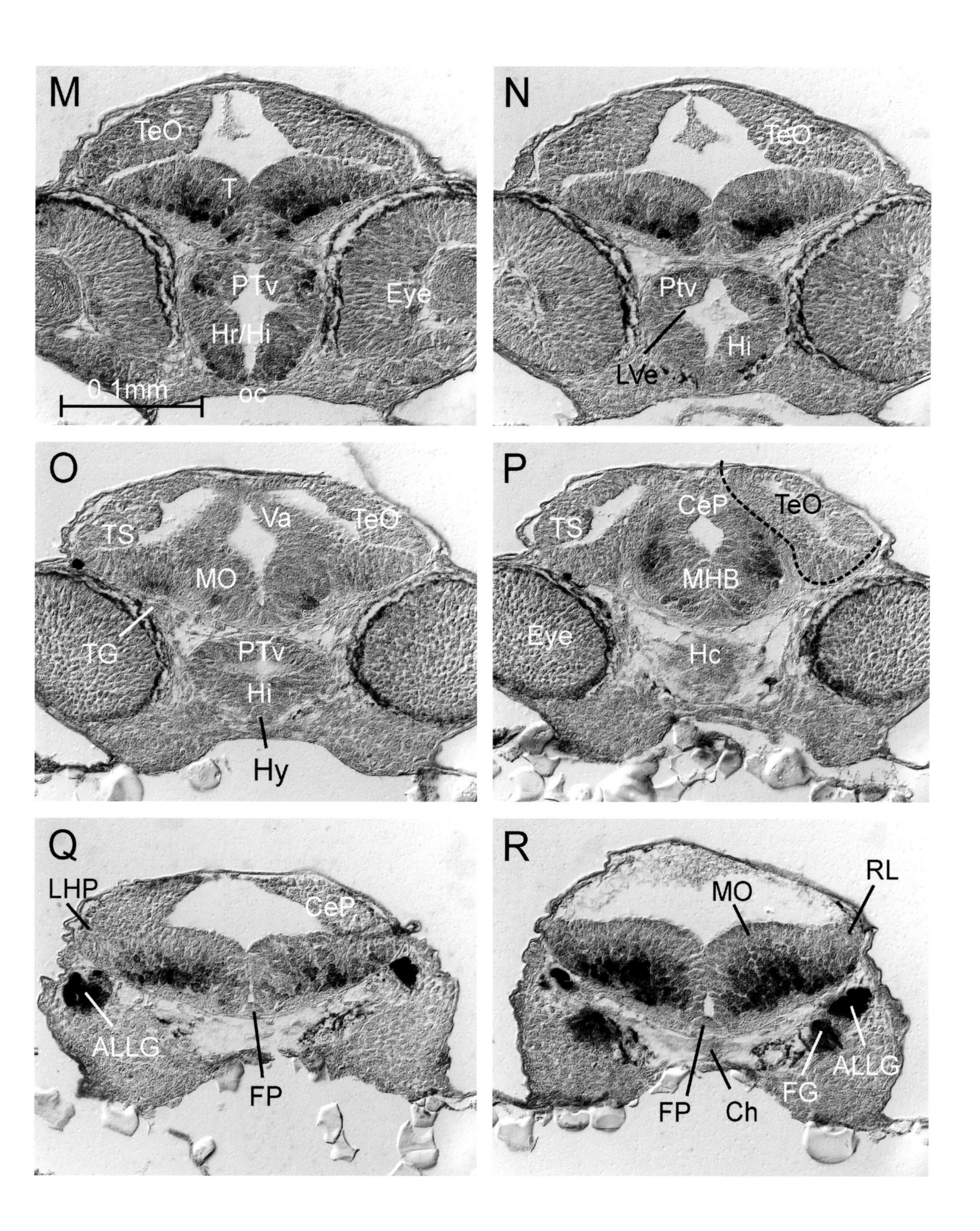
M
TeO
T
PTv
Eye
Hr/Hi
0.1mm
oc
N
TeO
Ptv
Hi
LVe
O
TS
Va
TeO
MO
TG
PTv
Hi
Hy
P
CeP
TeO
TS
MHB
Eye
Hc
Q
LHP
CeP
ALLG
FP
R
RL
MO
FP
Ch
FG
ALLG

Hu-proteins (continued, 4th plate)

Protein Expression Domains in the 2-Day Zebrafish Brain

Description of levels on facing page

Sensory organs/PNS: Patchy Hu signal in cellular lining of otic capsule and expression of Hu in various cranial nerve ganglia of peripheral nervous system.

CNS: The entire medulla oblongata from intermediate (characterized by otic capsule) up to caudal levels is strongly Hu positive in the periphery, but not the rhombic lip (rim of dorsal opening of medulla oblongata, the rhombic fossa). Note also that specialized floor plate midline cells are Hu negative.

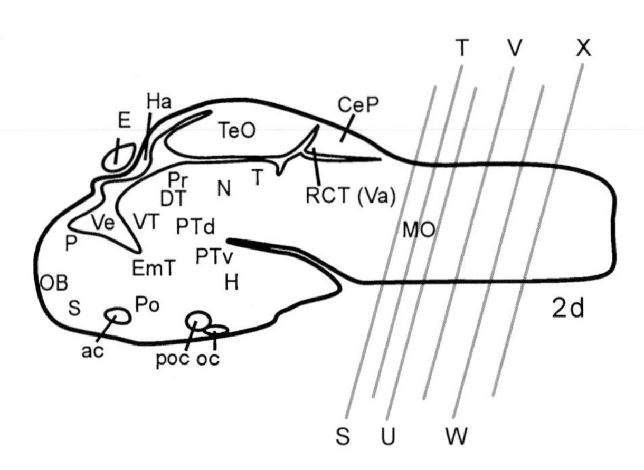

Abbreviations

ac	anterior commissure	P	pallium
CeP	cerebellar plate	Po	preoptic region
Ch	chorda dorsalis	poc	postoptic commissure
DT	dorsal thalamus (thalamus)	Pr	pretectum
E	epiphysis	PTd	dorsal part of posterior tuberculum
EmT	eminentia thalami	PTv	ventral part of posterior tuberculum
FP	floor plate		
GG	glossopharyngeal ganglion	RCT	rostral cerebellar thickening (valvula)
H	hypothalamus		
Ha	habenula	RL	rhombic lip
MO	medulla oblongata	S	subpallium
N	region of the nucleus of the medial longitudinal fascicle	T	midbrain tegmentum
		TeO	tectum opticum
OB	olfactory bulb	Va	valvula cerebelli
OC	otic capsule	Ve	brain ventricle
oc	optic chiasma	VG	vagal ganglion
OG	octaval ganglion	VT	ventral thalamus (prethalamus)

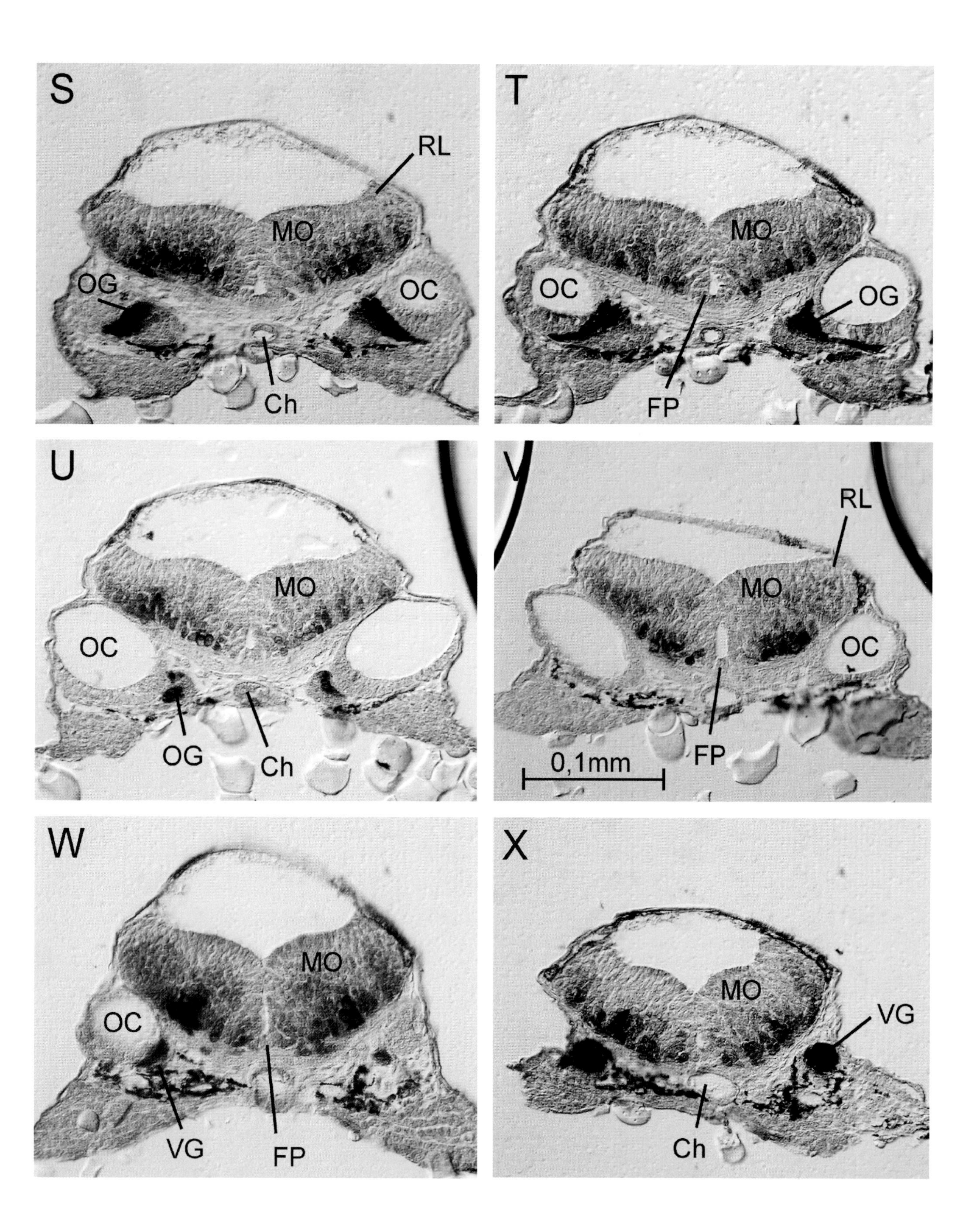
S
RL
MO
OG
OC
Ch
T
MO
OC
OG
FP
U
MO
OC
OC
OG
Ch
V
RL
MO
OC
FP
0,1mm
W
MO
OC
VG
FP
X
MO
VG
Ch

PCNA-protein (shown with monoclonal mouse antibody against the *p*roliferating *c*ell *n*uclear *a*ntigen; PC10 of Waseem and Lane 1990; source: DAKO, Code Nr. M 879, Glostrup, Denmark)
Description: Auxiliary protein of DNA polymerase-δ (Mathews et al. 1984). Expression in proliferating cells, especially during development.

Protein Expression Domains in the 3-Day Zebrafish Brain

General appearance: Consistent with a necessary role for mitosis, PCNA domains visualize patterns of CNS (and PNS) proliferation zones which typically lie directly at the ventricle along the entire neuraxis, generally medial to cells expressing neuronal determination (e.g., *neurod*) or differentiation (e.g., Hu) markers. Patterns of proliferative zones reveal prosomeric organization in posterior forebrain (i.e., pretectum, dorsal thalamus, ventral thalamus) and rhombomeric organization in medulla oblongata.

Description of levels on facing page

Sensory organs: Strong PCNA signal is restricted to retinal peripheral edge. Clusters of PCNA-positive cells are seen in olfactory epithelium.

CNS: In the telencephalon, pallial and subpallial PCNA domains, including domain in medial proliferative zone of olfactory bulb, lie directly at the ventricular surface at all anteroposterior levels; this also applies to eminentia thalami and preoptic region. The latter is separated from the anteriorly lying subpallium by the anterior commissure. There is no PCNA signal in M4, an early migrated cell aggregate possibly representing the future lateral nucleus of the ventral telencephalic area.

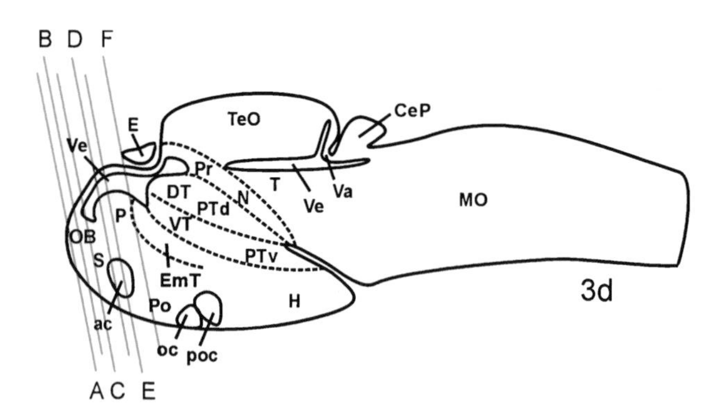

Abbreviations

ac	anterior commissure	Pi	pigment
CeP	cerebellar plate	Po	preoptic region
DT	dorsal thalamus (thalamus)	poc	postoptic commissure
E	epiphysis	Pr	pretectum
EmT	eminentia thalami	PTd	dorsal part of posterior tuberculum
EPi	eye pigment	PTv	ventral part of posterior tuberculum
H	hypothalamus	S	subpallium
Ha	habenula	Sd	dorsal division of S
lfb	lateral forebrain bundle	Sv	ventral division of S
M4	migrated telencephalic cells	T	midbrain tegmentum
MO	medulla oblongata	TeO	tectum opticum
N	region of the nucleus of the medial longitudinal fascicle	TVe	telencephalic ventricle
OB	olfactory bulb	Va	valvula cerebelli
oc	optic chiasma	Ve	brain ventricle
OE	olfactory epithelium	VT	ventral thalamus (prethalamus)
P	pallium		

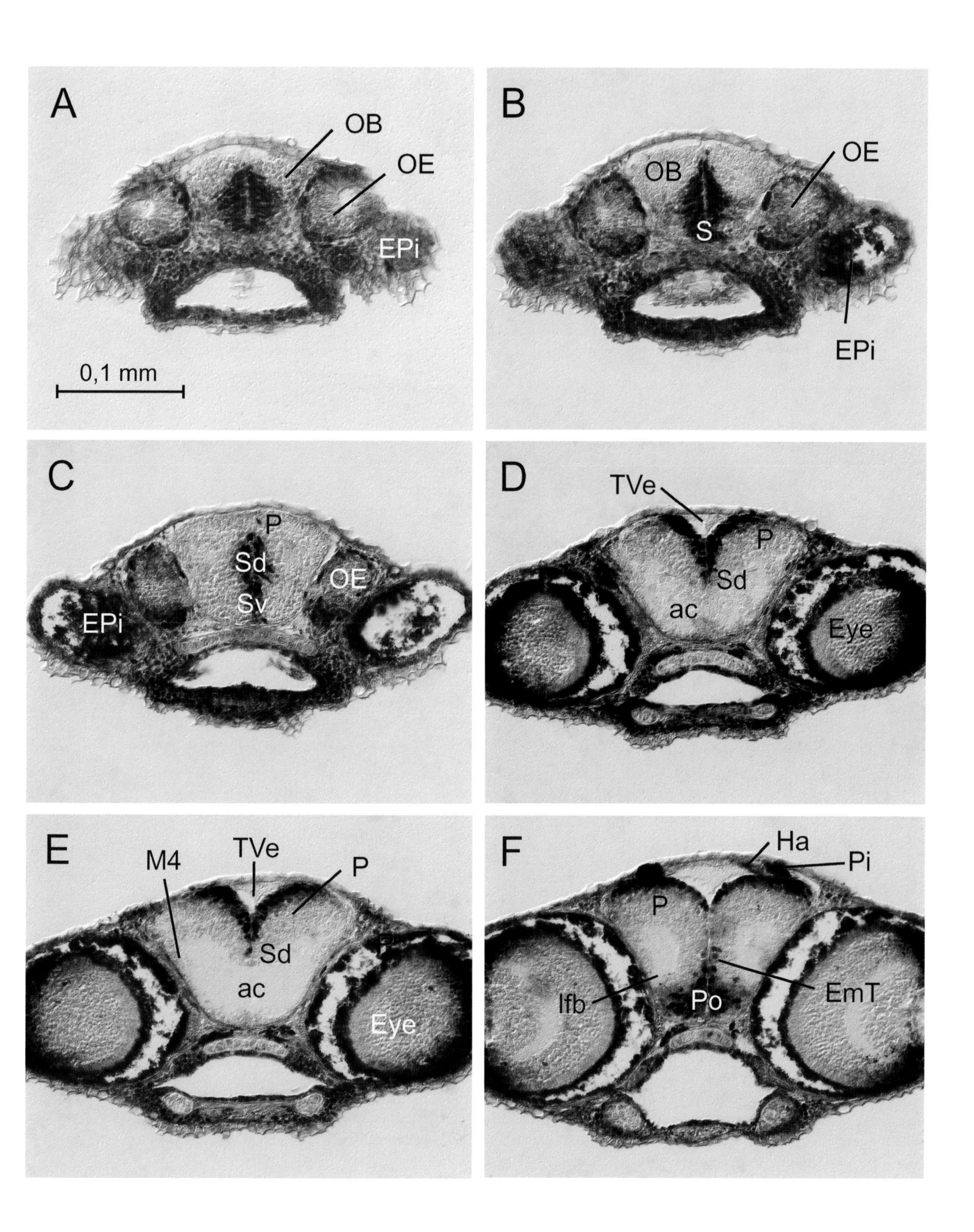
A
OB
OE
EPi
0,1 mm
B
OB
OE
S
EPi
C
P
Sd
Sv
OE
EPi
D
TVe
P
Sd
ac
Eye
E
M4
TVe
P
Sd
ac
Eye
F
Ha
Pi
P
EmT
lfb
Po

PCNA-protein (continued, 2nd plate)

Protein Expression Domains in the 3-Day Zebrafish Brain

Description of levels on facing page

Sensory organs: Strong PCNA signal is restricted to retinal peripheral edge.

CNS: At these levels, strong and extensive PCNA signals are present directly at the respective ventricular lining in the preoptic region (which is separated caudally by the postoptic commissure from the anterior part of the emerging hypothalamus), the rostral, as well as the intermediate hypothalami (the latter is characterized by the lateral ventricular recess). In the diencephalon, such ventricular PCNA domains are present in habenula, dorsal thalamus proper, ventral thalamus (extends particularly far laterally) and pretectum, as well as in the posterior tuberculum (dorsal and ventral divisions visible on facing page). Within the optic tectum, medial (dorsal and ventral parts visible rostrally) and lateral PCNA domains emerge (note the PCNA-negative midline area, likely representing torus longitudinalis). Except for a small ventricular spot, there is no PCNA expression in the zona limitans intrathalamica and none in the epiphysis. Note cluster of PCNA-positive cells in the far migrated posterior tubercular region M2 (future preglomerular complex).

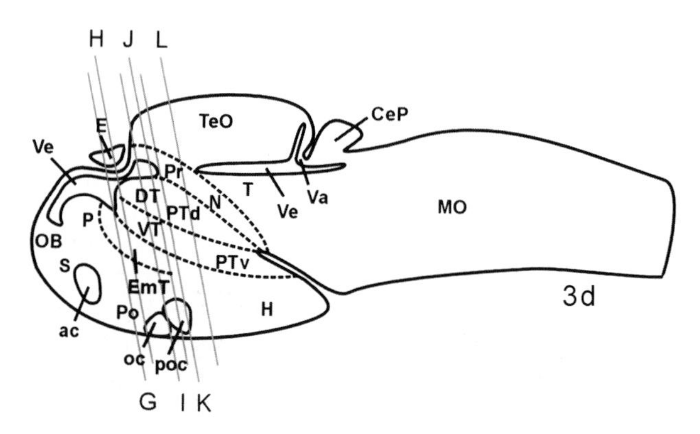

Abbreviations

ac	anterior commissure	OB	olfactory bulb
CeP	cerebellar plate	oc	optic chiasma
DT	dorsal thalamus (thalamus)	on	optic nerve
E	epiphysis	P	pallium
EmT	eminentia thalami	pc	posterior commissure
fr	fasciculus retroflexus	Po	preoptic region
H	hypothalamus	poc	postoptic commissure
Ha	habenula	Pr	pretectum
Hi	intermediate hypothalamus	PT	posterior tuberculum
Hr	rostral hypothalamus	PTd	dorsal part of posterior tuberculum
l	lateral tectal proliferation zone	PTv	ventral part of posterior tuberculum
lfb	lateral forebrain bundle	S	subpallium
LVe	lateral recess ventricle of H	SCO	subcommissural organ
M1	migrated pretectal area	T	midbrain tegmentum
M2	migrated posterior tubercular area	TeO	tectum opticum
M3	migrated area of EmT	TL	torus longitudinalis
m	medial tectal proliferation zone	Va	valvula
MO	medulla oblongata	Ve	brain ventricle
N	region of the nucleus of medial longitudinal fascicle	VT	ventral thalamus (prethalamus)
		ZLI	zona limitans intrathalamica

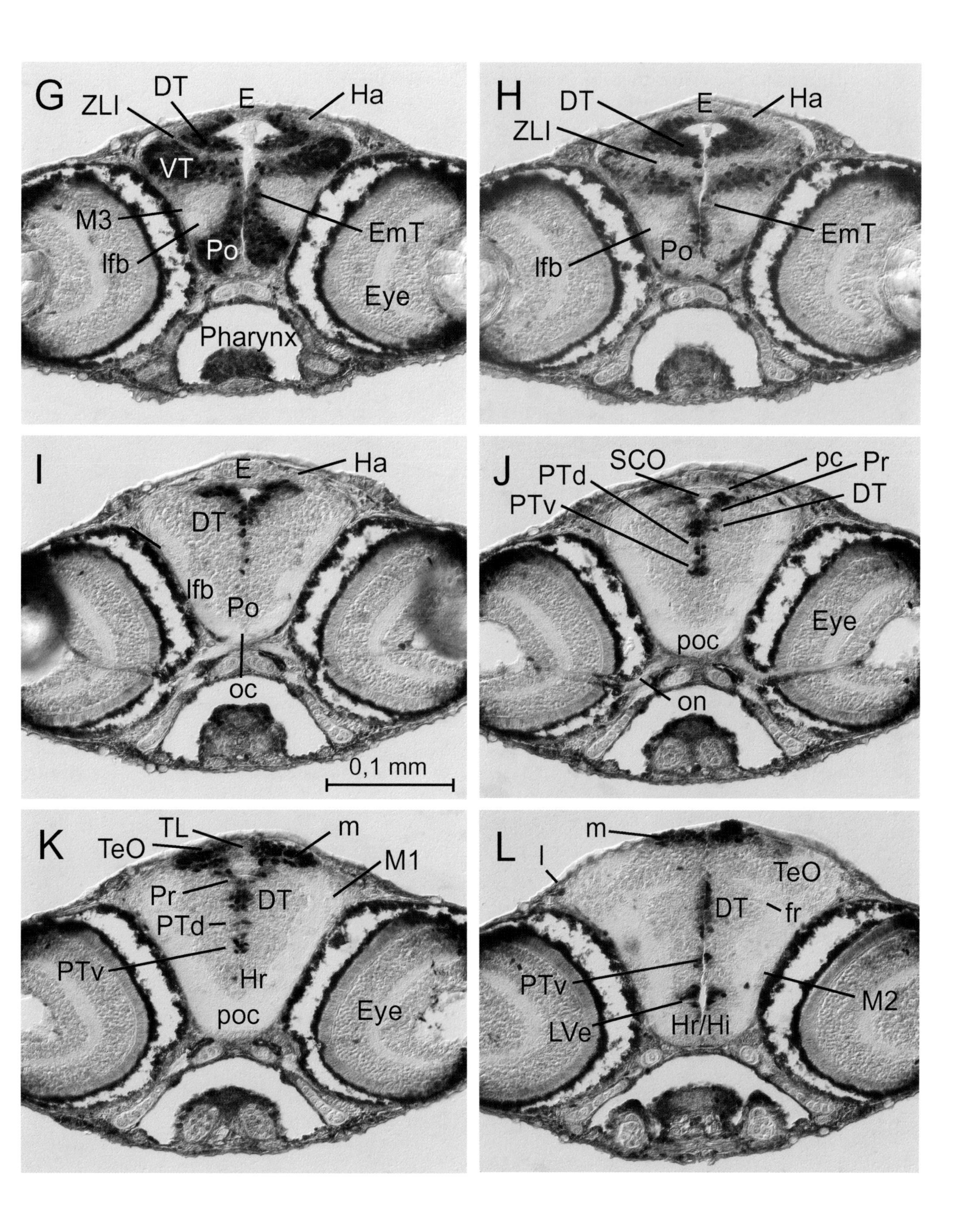
G
ZLI
DT
E
Ha
VT
M3
lfb
Po
EmT
Eye
Pharynx
H
DT
E
Ha
ZLI
lfb
Po
EmT
I
E
Ha
DT
lfb
Po
oc
0,1 mm
J
PTd
SCO
pc
Pr
PTv
DT
Eye
poc
on
K
TL
m
TeO
M1
Pr
DT
PTd
PTv
Hr
poc
Eye
L
m
l
TeO
DT
fr
PTv
M2
LVe
Hr/Hi

PCNA-protein (continued, 3rd plate)

Protein Expression Domains in the 3-Day Zebrafish Brain

Description of levels on facing page

Sensory organs/PNS: Strong PCNA signal is restricted to retinal peripheral edge. Patchy PCNA expression is also present in various cranial nerve ganglia of peripheral nervous system.

CNS: Strong ventricular PCNA signal is present in intermediate and caudal hypothalami (the latter is characterized by the posterior ventricular recess). The hypothalamus extends into the ventral part of the inferior lobe and is interpreted as basal plate of anterior forebrain (secondary prosencephalon). The proliferation of the ventral posterior tuberculum continues to extend into the dorsal portion of the inferior lobe (interpreted as basal plate of ventral thalamic prosomere/P3). Hypophysis shows a few PCNA-positive cells. Also, very few PCNA-positive cells are seen in the synencephalic region of the nucleus of the medial longitudinal fascicle (N; basal plate of P1), which is replaced more caudally by a distinct PCNA-positive blob of the midbrain tegmentum. Medial and lateral PCNA domains of optic tectum continue to be present at these levels. Also the (medullary) midbrain–hindbrain boundary shows a strong PCNA signal. Anterior medulla oblongata contains patchy PCNA cell clusters in the dorsal midline and along the ventral midline ventricle.

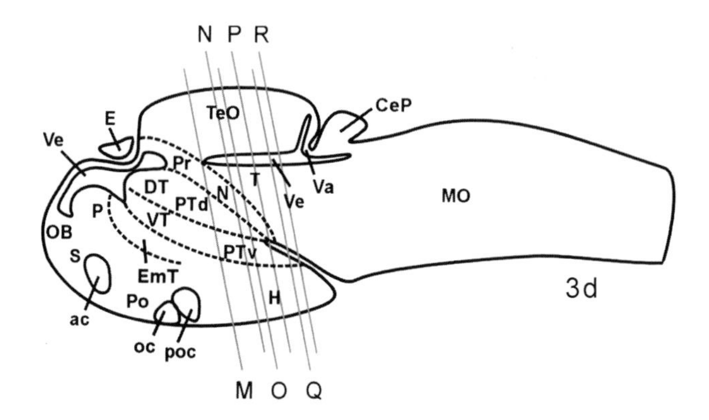

Abbreviations

ac	anterior commissure	NIn	nucleus interpeduncularis
ALLG	anterior lateral line ganglion	OB	olfactory bulb
CeP	cerebellar plate	oc	optic chiasma
Ch	chorda dorsalis	OG	octaval ganglion
DT	dorsal thalamus (thalamus)	P	pallium
E	epiphysis	Po	preoptic region
EmT	eminentia thalami	Poc	postoptic commissure
EPi	eye pigment	Pr	pretectum
H	hypothalamus	PTd	dorsal part of posterior tuberculum
Hc	caudal hypothalamus		
Hi	intermediate hypothalamus	PTv	ventral part of posterior tuberculum
Hy	hypophysis (pituitary)		
l	lateral optic tectum	PVe	posterior ventricular recess of H
LVe	lateral ventricular recess of H		
M2	migrated posterior tubercular area	S	subpallium
		T	midbrain tegmentum
m	medial optic tectum	TeO	tectum opticum
MHB	midbrain–hindbrain boundary	TG	trigeminal ganglion
MO	medulla oblongata	Va	valvula cerebelli
N	region of the nucleus of medial longitudinal fascicle	Ve	brain ventricle
		VT	ventral thalamus (prethalamus)

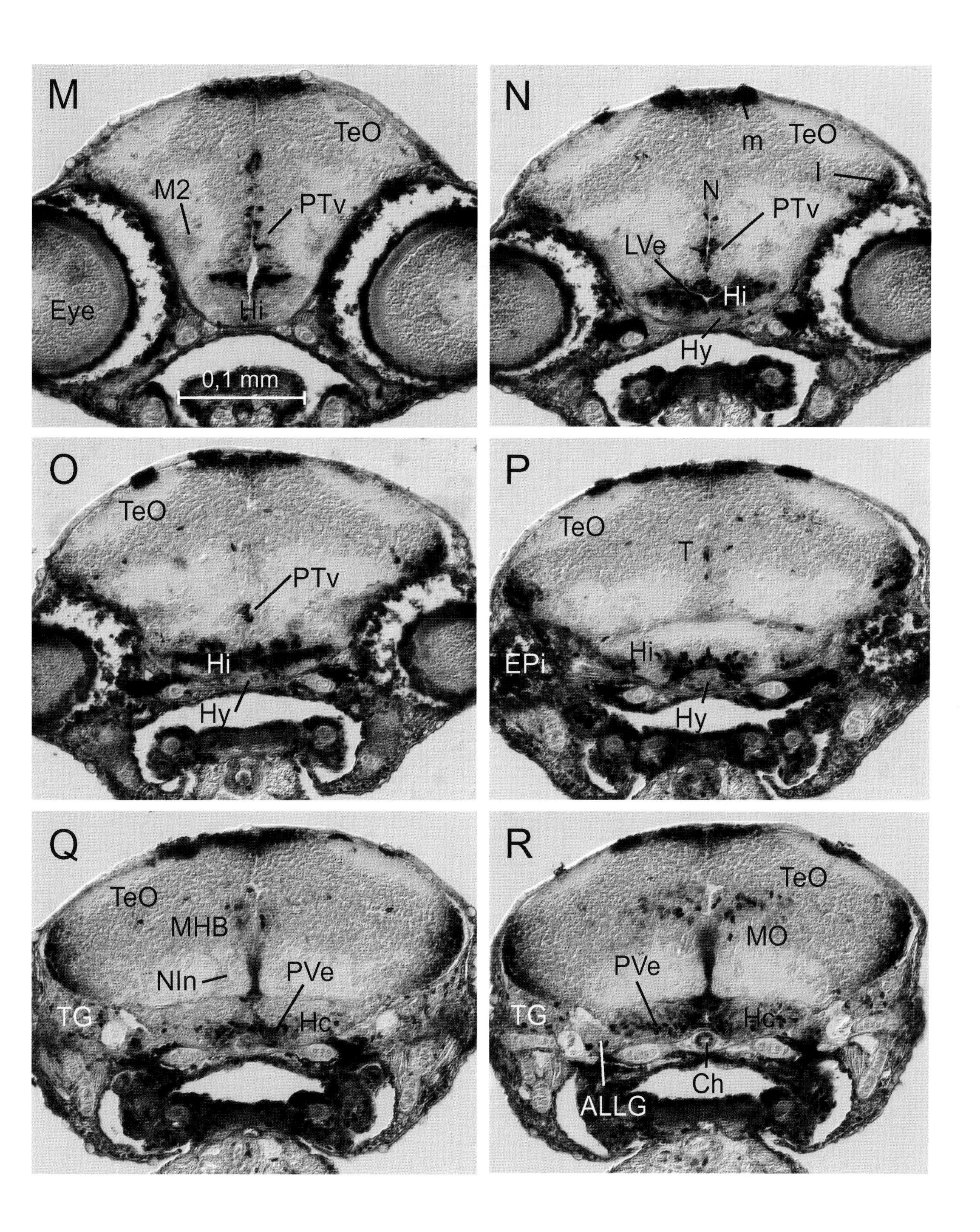
M
TeO
M2
PTv
Eye
Hi
0,1 mm
N
m
TeO
I
N
PTv
LVe
Hi
Hy
O
TeO
PTv
Hi
Hy
P
TeO
T
EPi
Hi
Hy
Q
TeO
MHB
NIn
PVe
TG
Hc
R
TeO
MO
PVe
TG
Hc
Ch
ALLG

PCNA-protein (continued, 4th plate)

Protein Expression Domains in 3-Day Zebrafish Brain

Description of levels on facing page

Sensory organs/PNS: Patchy PCNA signal in cellular lining of otic capsule and in various cranial nerve ganglia of peripheral nervous system.

CNS: At these levels, PCNA-positive cells are seen in the torus semicircularis, which lies in the caudolateral midbrain roof. Medial, basal and lateral PCNA domains of optic tectum increase in extent caudally (note that PCNA cells designated as b partly lie ventral to tectal ventricle and, thus, belong to medulla oblongata). The most caudal tectal PCNA cells are seen to lie on top of the emerging cerebellar plate. The latter displays PCNA-positive cells in the ventral proliferative layer as well as in the external granular layer. There are also distinct PCNA domains in valvula cerebelli and in eminentia granularis. The rhombic lip (rim of dorsal opening of medulla oblongata, the rhombic fossa) is strongly PCNA positive. In contrast, the remainder of the medulla oblongata shows a rather weak PCNA signal at these levels, mostly along the ventral midline ventricle.

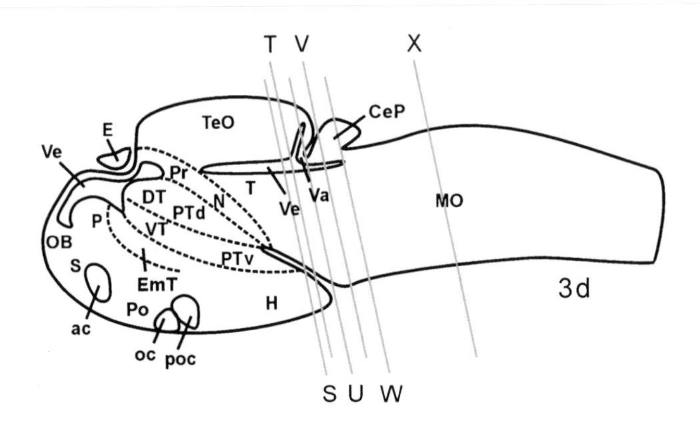

Abbreviations

ac	anterior commissure
b	basal optic tectum
CeP	cerebellar plate
Ch	chorda dorsalis
DT	dorsal thalamus (thalamus)
E	epiphysis
EG	eminentia granularis
EGL	external granular layer
EmT	eminentia thalami
H	hypothalamus
Hc	caudal hypothalamus
l	lateral optic tectum
m	medial optic tectum
MO	medulla oblongata
N	region of the nucleus of medial longitudinal fascicle
OB	olfactory bulb
OC	otic capsule
oc	optic chiasma
OG	octaval ganglion
P	pallium
Pi	pigment
PLLG	posterior lateral line ganglion
Po	preoptic region
poc	postoptic commissure
Pr	pretectum
PTd	dorsal part of posterior tuberculum
PTv	ventral part of posterior tuberculum
PVe	posterior recess ventricle of H
RL	rhombic lip
S	subpallium
T	midbrain tegmentum
TeO	tectum opticum
TeVe	tectal ventricle
TS	torus semicircularis
Va	valvula cerebelli
VCP	ventral cerebellar proliferation
Ve	brain ventricle
VG	vagal ganglion
VT	ventral thalamus (prethalamus)

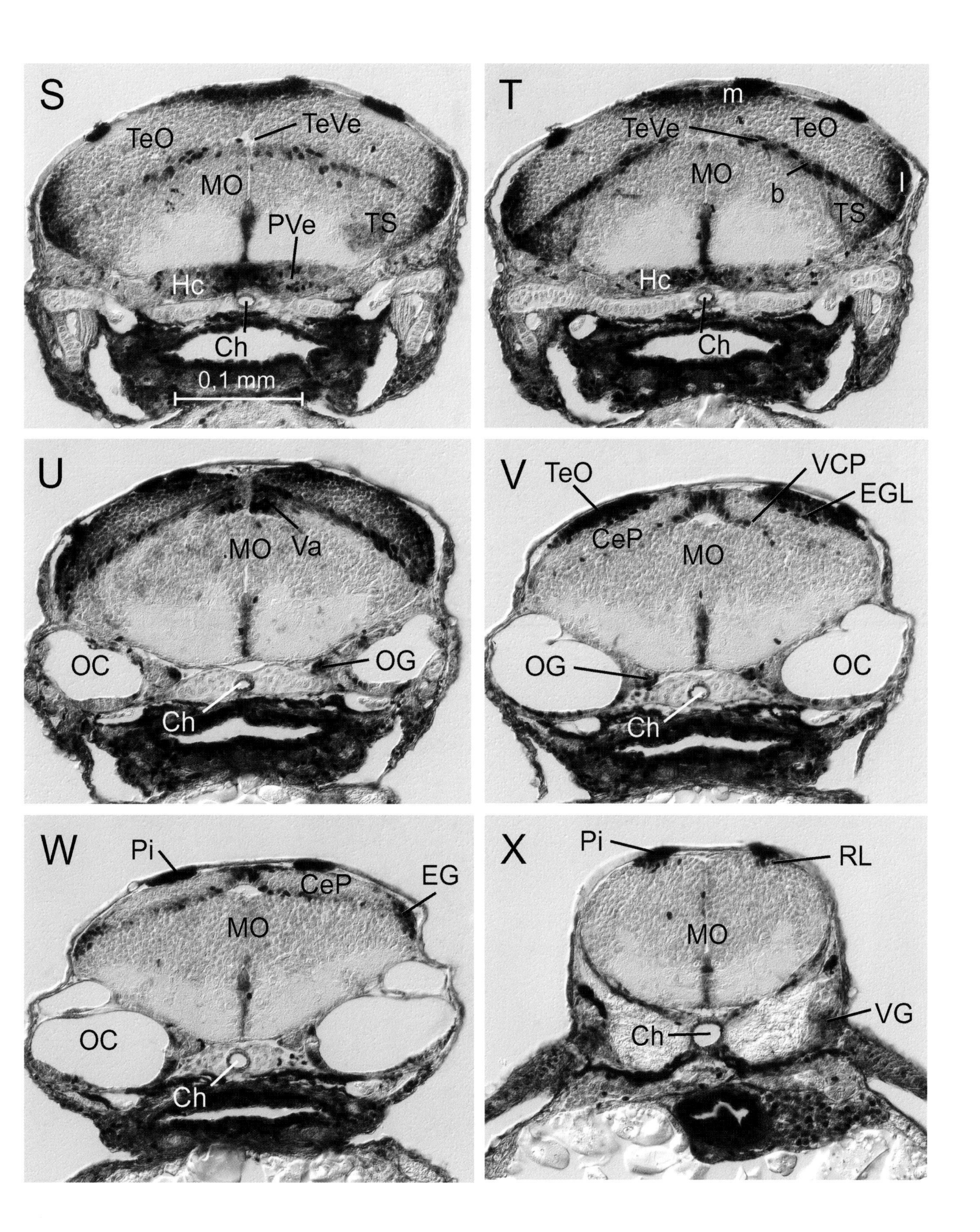
S
TeO
TeVe
MO
PVe
TS
Hc
Ch
0,1 mm
T
m
TeVe
TeO
MO
b
l
TS
Hc
Ch
U
MO
Va
OC
OG
Ch
V
TeO
VCP
EGL
CeP
MO
OG
OC
Ch
W
Pi
EG
CeP
MO
OC
Ch
X
Pi
RL
MO
VG
Ch

neurod (ZFIN ID: ZDB-GENE-990415-172; previous names: Beta2/NeuroD; nrd; neuro-D)
Description: Proneural gene, codes for basic Helix–Loop–Helix transcription factor, expressed in early determined neuronal and sensory cells.

Gene Expression Domains in the 3-Day Zebrafish Brain

General appearance: Consistent with a role in early neuronal determination, *neurod* expression domains lie one to several cell rows away from the brain ventricle, flanked by both *neurod*-negative proliferation zones towards the ventricle (compare with PCNA) and more peripheral differentiating cell masses towards the pia (compare with Hu). Although *neurod* expression patterns remain qualitatively similar to 2 days, some domains start to thin out or are less extensive. The subpallium, ventral thalamus, preoptic region and hypothalamus still show complete absence of *neurod* expression.

Description of levels on facing page

Sensory organs: Strong *neurod* signal in outer and inner nuclear layers of central retina, but not in ganglionic layer and retinal peripheral edge. Small clusters of *neurod*-positive cells are present in the olfactory epithelium.

CNS: In the telencephalon, the large pallial domain lies somewhat remote from the ventricular surface. The *neurod*-positive cells in the olfactory bulb sit towards its periphery and continue posteriorly to lie laterally to the subpallium (*neurod* negative). Note weakly stained *neurod* cells in M4, an early migrated cell aggregate possibly representing the future lateral nucleus of the ventral telencephalic area. In the diencephalon, sizable blobs of expression are present in epiphysis, habenula and dorsal thalamus proper. Caudal to the pallium, the *neurod* domain of the eminentia thalami appears. The subpallium (*neurod* negative) is caudally replaced by the preoptic region (*neurod* negative), separated from it by the anterior commissure. The anterior optic tectum contains a large medial *neurod* domain. There is no *neurod* expression in subpallium, preoptic region and ventral thalamus.

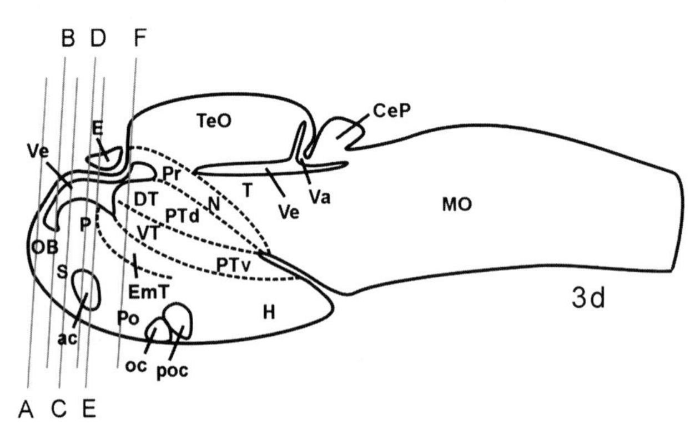

Abbreviations

ac	anterior commissure	P	pallium
CeP	cerebellar plate	Pi	pigment
DT	dorsal thalamus (thalamus)	Po	preoptic region
DVe	diencephalic ventricle	poc	postoptic commissure
E	epiphysis	Pr	pretectum
EmT	eminentia thalami	PTd	dorsal part of posterior tuberculum
EPi	eye pigment	PTv	ventral part of posterior tuberculum
H	hypothalamus	S	subpallium
Ha	habenula	Sd	dorsal division of S
lfb	lateral forebrain bundle	Sv	ventral division of S
M4	telencephalic migrated area	T	midbrain tegmentum
MO	medulla oblongata	TeO	tectum opticum
N	region of the nucleus of medial longitudinal fascicle	TVe	telencephalic ventricle
OB	olfactory bulb	Va	valvula cerebelli
oc	optic chiasma	Ve	brain ventricle
OE	olfactory epithelium	VT	ventral thalamus (prethalamus)

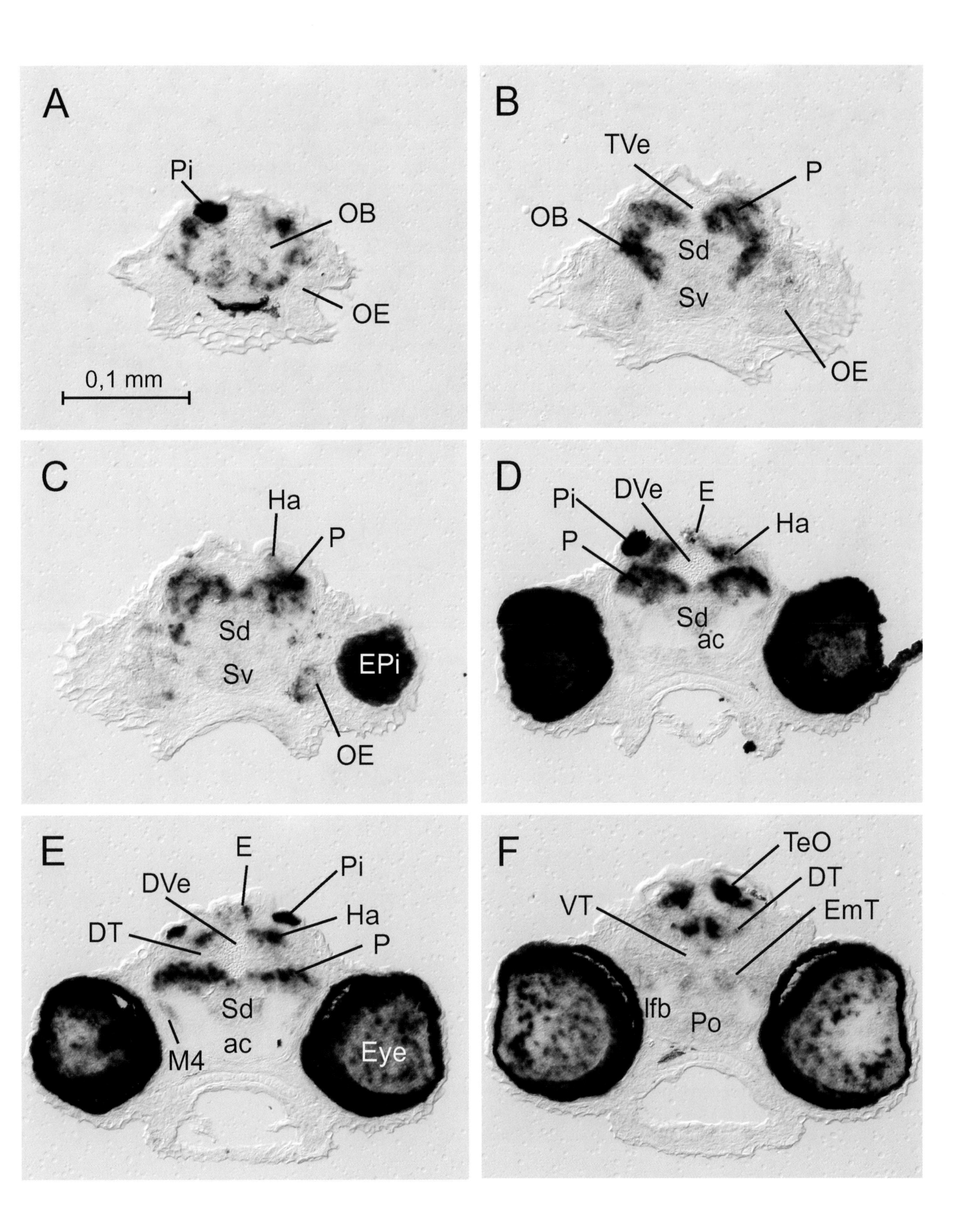
A
Pi
OB
OE
0,1 mm
B
TVe
P
OB
Sd
Sv
OE
C
Ha
P
Sd
Sv
EPi
OE
D
Pi
DVe
E
Ha
P
Sd
ac
E
E
Pi
DVe
Ha
DT
P
Sd
ac
M4
Eye
F
TeO
DT
VT
EmT
lfb
Po

neurod (continued, 2nd plate)

Gene Expression Domains in the 3-Day Zebrafish Brain

Description of levels on facing page

Sensory organs: Strong *neurod* signal in outer and inner nuclear layers of central retina, but not in ganglionic layer and retinal peripheral edge.

CNS: At these levels, distinct *neurod* domain of the eminentia thalami apparently arches around the lateral forebrain bundle. However, the most peripherally migrated cells in this region (compare with Hu), lying in the position of the future entopeduncular complex (M3), are *neurod* negative (as are the far migrated pretectal cells, future superficial pretectum; M1). The postoptic commissure separates the preoptic region caudally from the anterior part of the emerging hypothalamus (*neurod* negative). The diencephalon displays a large blob of *neurod* expression in the dorsal thalamus proper and smaller ones in the pretectum and the posterior tuberculum (domain in ventral division more prominent than in dorsal division). The migrated posterior tubercular region (future preglomerular region; M2) contains distinct, strongly *neurod*-positive cell clusters. Caudal to the posterior tuberculum, there is now only very weak *neurod* staining in the basal synencephalon, i.e., the region of the nucleus of the medial longitudinal fascicle. In the optic tectum, medial, basal (note that these basal *neurod* cells may partly lie ventral to tectal ventricle and, thus, belong to tegmentum) and lateral *neurod* domains are present. At most anterior tectal levels, there is strong *neurod* staining in the presumptive torus longitudinalis; the large domain below the anterior tectal ventricle may belong to the griseum tectale. There is no *neurod* expression in preoptic region, ventral thalamus and hypothalamus.

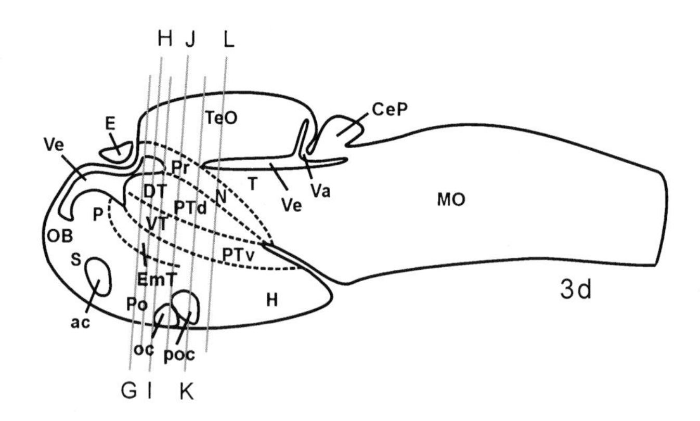

Abbreviations

ac	anterior commissure	OB	olfactory bulb
CeP	cerebellar plate	oc	optic chiasma
DT	dorsal thalamus (thalamus)	P	pallium
E	epiphysis	Pi	pigment
EmT	eminentia thalami	Po	preoptic region
GT	griseum tectale	poc	postoptic commissure
H	hypothalamus	Pr	pretectum
Hi	intermediate hypothalamus	PTd	dorsal part of posterior tuberculum
Hr	rostral hypothalamus	PTv	ventral part of posterior tuberculum
lfb	lateral forebrain bundle	S	subpallium
M1	migrated pretectal area	T	midbrain tegmentum
M2	migrated posterior tubercular area	TeO	tectum opticum
M3	migrated area of EmT	TL	torus longitudinalis
MO	medulla oblongata	Va	valvula
N	region of the nucleus of the medial longitudinal fascicle	Ve	brain ventricle
		VT	ventral thalamus (prethalamus)

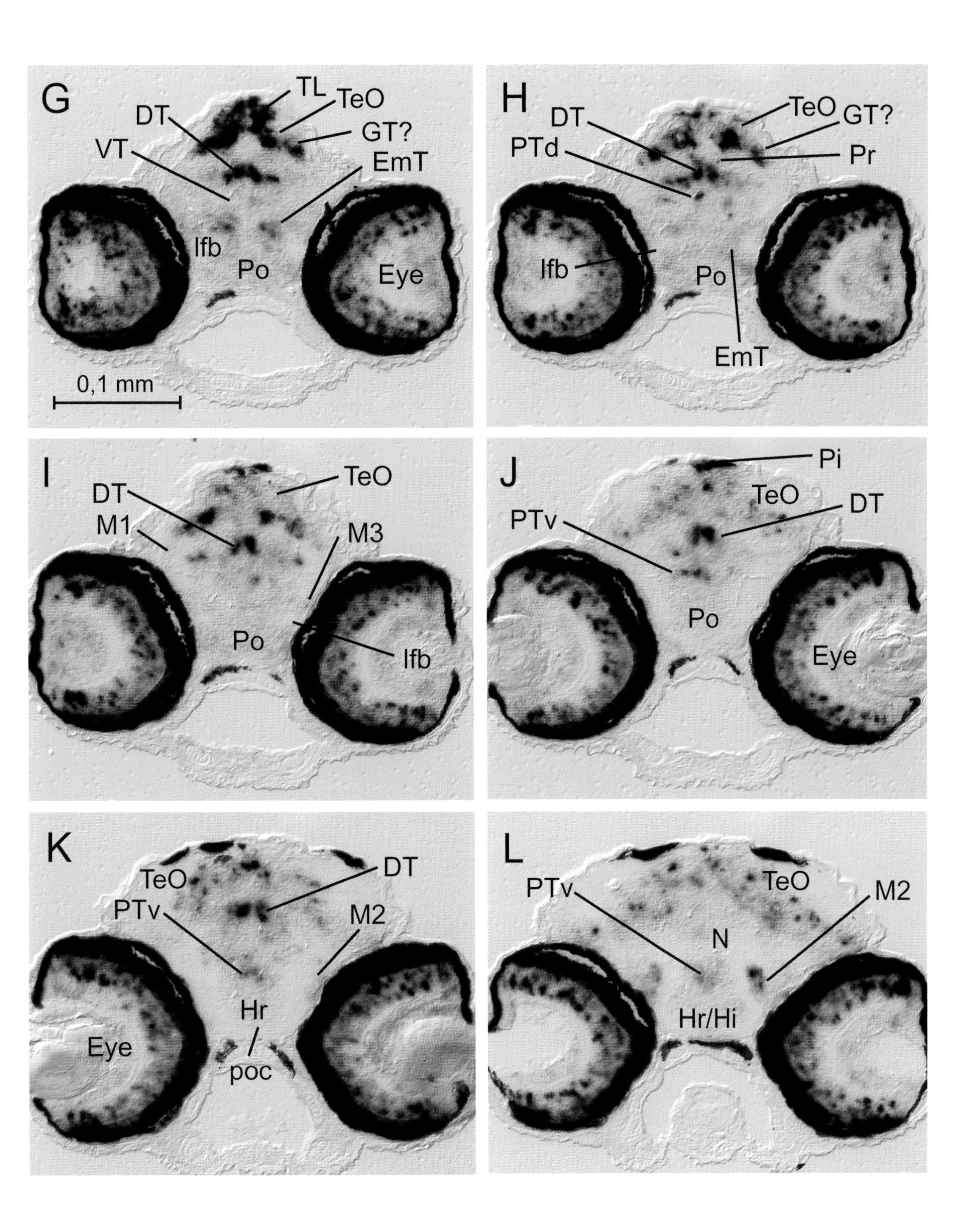
G
TL
TeO
DT
GT?
VT
EmT
lfb
Po
Eye
0,1 mm
H
DT
TeO
GT?
PTd
Pr
lfb
Po
EmT
I
TeO
DT
M1
M3
Po
lfb
J
Pi
TeO
DT
PTv
Po
Eye
K
DT
TeO
PTv
M2
Hr
Eye
poc
L
PTv
TeO
M2
N
Hr/Hi

neurod (continued, 3rd plate)

Gene Expression Domains in the 3-Day Zebrafish Brain

Description of levels on facing page

Sensory organs/PNS: Strong *neurod* signal in outer and inner nuclear layers of central retina, but not in ganglionic layer and retinal peripheral edge. Expression of *neurod* is also present in cranial nerve ganglia of peripheral nervous system.

CNS: Dual composition of inferior lobe is still apparent: the ventral, hypothalamic part (interpreted as basal plate of anterior forebrain/secondary prosencephalon) remains *neurod* negative, whereas the dorsal inferior lobe portion is *neurod* positive (interpreted as basal plate of ventral thalamic prosomere/P3). Hypophysis continues to show a *neurod* signal. The midbrain tegmentum *neurod* domain is still distinct, if much weaker, and fades out towards caudally adjacent, strongly proliferative (*neurod* negative) (medullary) midbrain–hindbrain boundary. Large medial, basal (note that these basal *neurod* cells may partly lie ventral to tectal ventricle and, thus, belong to medulla oblongata) and lateral *neurod* domains of optic tectum continue to be present at these levels, as is the *neurod* domain in the torus semicircularis. The cerebellar plate displays now a basal and a peripheral (external granular layer) *neurod*-positive layer and the valvula cerebelli is additionally *neurod* positive (in contrast to 2 days). Also, the eminentia granularis is now clearly recognizable by exhibiting a strong *neurod* signal. Anterior medulla oblongata contains a much more diffusely distributed *neurod* signal compared with 2 days. There is still no *neurod* expression in hypothalamus.

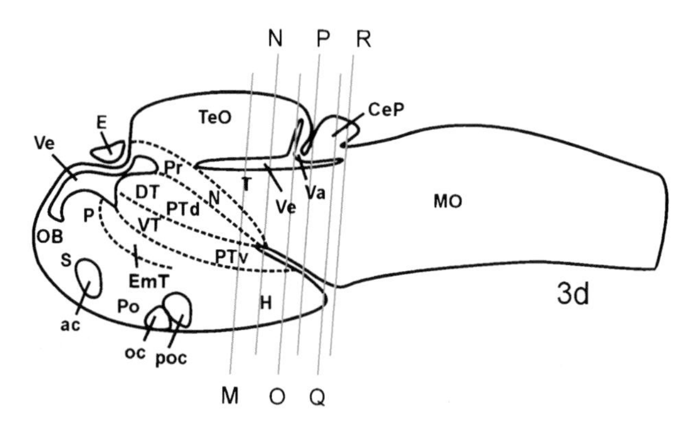

Abbreviations

ac	anterior commissure	OB	olfactory bulb
ALLG	anterior lateral line ganglion	oc	optic chiasma
CeP	cerebellar plate	P	pallium
Ch	chorda dorsalis	Po	preoptic region
DT	dorsal thalamus (thalamus)	poc	postoptic commissure
E	epiphysis	Pr	pretectum
EG	eminentia granularis	PTd	dorsal part of posterior tuberculum
EGL	external granular layer		
EmT	eminentia thalami	PTv	ventral part of posterior tuberculum
EPi	eye pigment		
H	hypothalamus	S	subpallium
Hc	caudal hypothalamus	T	midbrain tegmentum
Hi	intermediate hypothalamus	TeO	tectum opticum
Hr	rostral hypothalamus	TG	trigeminal ganglion
Hy	hypophysis (pituitary)	TS	torus semicircularis
M2	migrated posterior tubercular area	Va	valvula cerebelli
MHB	midbrain–hindbrain boundary	VCP	ventral cerebellar proliferation
MO	medulla oblongata	Ve	brain ventricle
N	region of the nucleus of the medial longitudinal fascicle	VT	ventral thalamus (prethalamus)

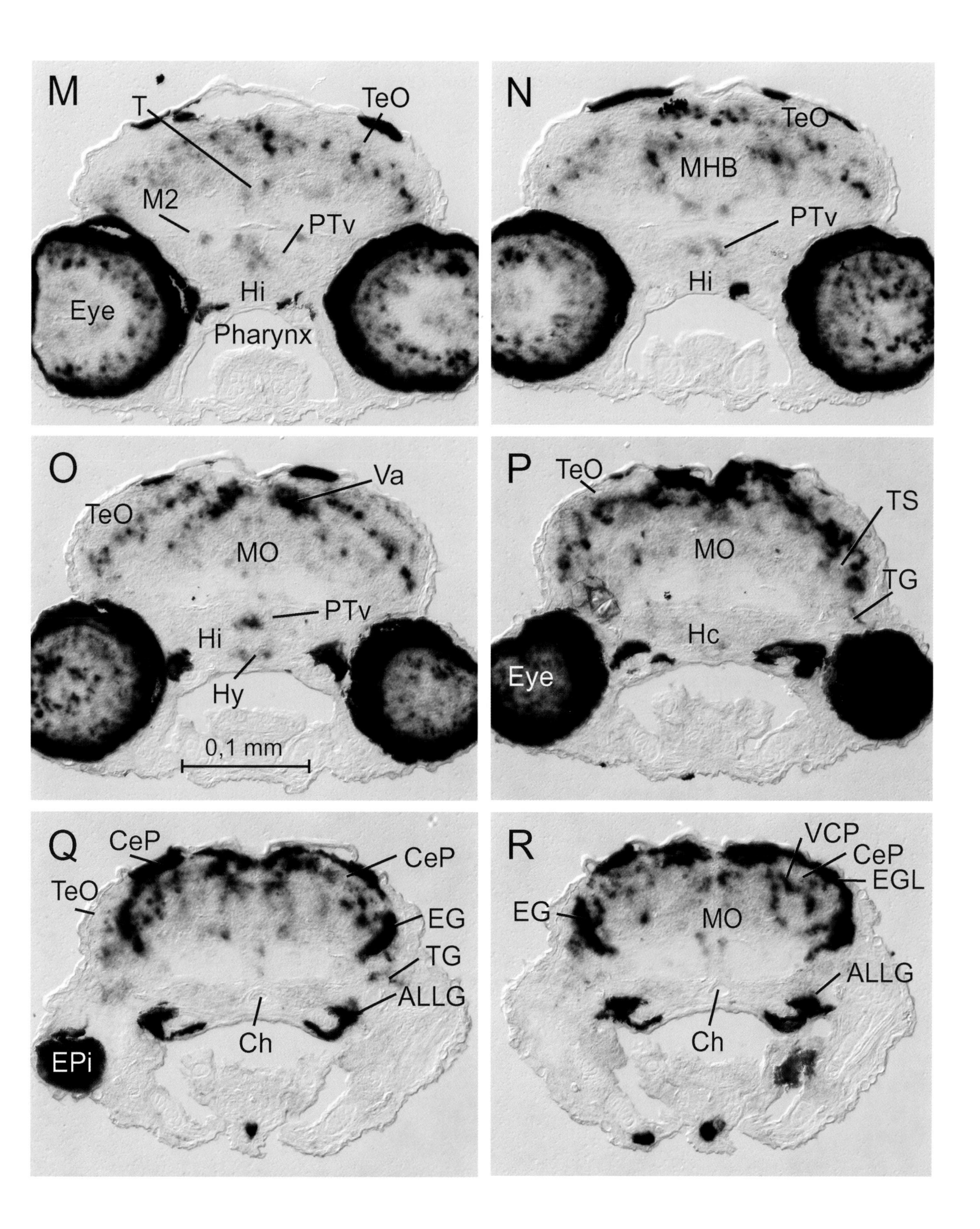
M
T
TeO
M2
PTv
Hi
Eye
Pharynx
N
TeO
MHB
PTv
Hi
O
Va
TeO
MO
PTv
Hi
Hy
0,1 mm
P
TeO
TS
MO
TG
Hc
Eye
Q
CeP
CeP
TeO
EG
TG
ALLG
Ch
EPi
R
VCP
CeP
EGL
EG
MO
ALLG
Ch

neurod (continued, 4th plate)

Gene Expression Domains in the 3-Day Zebrafish Brain

Description of levels on facing page

Sensory organs/PNS: Expression of *neurod* in various cranial nerve ganglia of peripheral nervous system.

CNS: At these medullary levels, the rhombic lip (rim of dorsal opening of medulla oblongata, the rhombic fossa) appears largely *neurod* negative (in contrast to 2 days), but many *neurod*-expressing cells still emanate from it ventrally at the lateral periphery of the rhombencephalon. Approaching the obex (boundary to spinal cord), the rhombic lip fuses medially, closing the rhombic fossa. At more caudal levels, the *neurod* signal in the midline medulla oblongata becomes stronger than anteriorly, with various distinct clusters of *neurod*-positive cells emanating ventrally into the rhombencephalic gray matter.

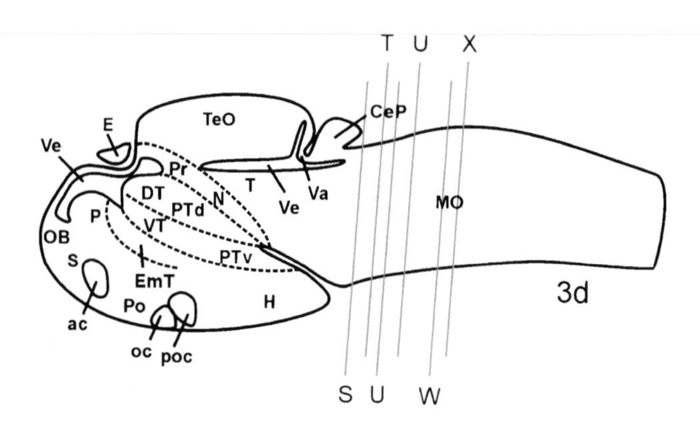

Abbreviations

ac	anterior commissure	P	pallium
CeP	cerebellar plate	Pi	pigment
Ch	chorda dorsalis	PLLG	posterior lateral line ganglion
DT	dorsal thalamus (thalamus)	Po	preoptic region
E	epiphysis	poc	postoptic commissure
EmT	eminentia thalami	Pr	pretectum
GG	glossopharyngeal ganglion	PTd	dorsal part of posterior tuberculum
H	hypothalamus	PTv	ventral part of posterior tuberculum
LHP	lateral hinge-point between cerebellum/medulla oblongata	RL	rhombic lip
MO	medulla oblongata	S	subpallium
N	region of the nucleus of the medial longitudinal fascicle	T	midbrain tegmentum
OB	olfactory bulb	TeO	tectum opticum
OC	otic capsule	Va	valvula cerebelli
oc	optic chiasma	Ve	brain ventricle
OG	octaval ganglion	VG	vagal ganglion
		VT	ventral thalamus (prethalamus)

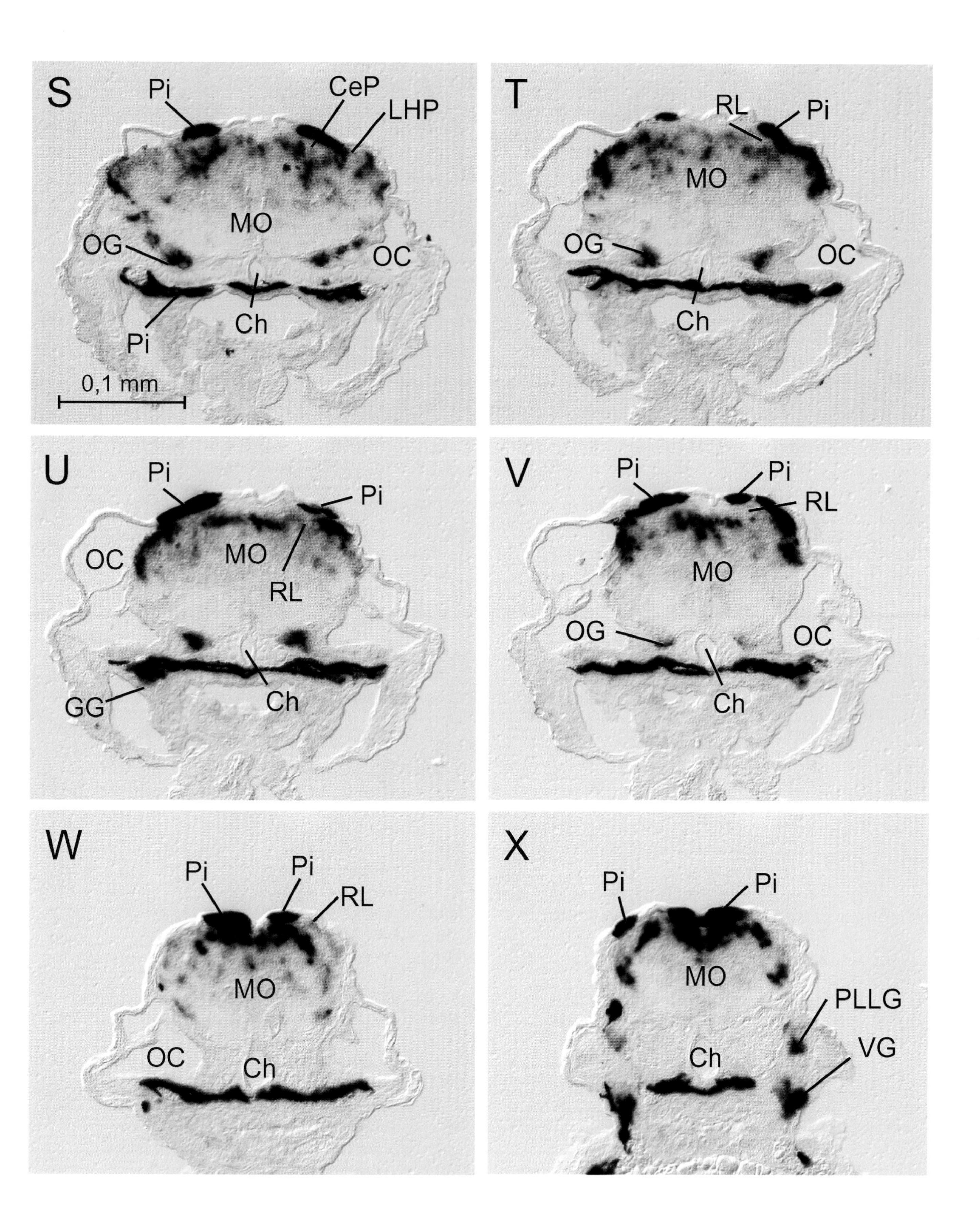
S
Pi
CeP
LHP
MO
OG
OC
Pi
Ch
0,1 mm
T
RL
Pi
MO
OG
OC
Ch
U
Pi
Pi
OC
MO
RL
GG
Ch
V
Pi
Pi
RL
MO
OG
OC
Ch
W
Pi
Pi
RL
MO
OC
Ch
X
Pi
Pi
MO
PLLG
Ch
VG

Hu-proteins (shown with anti-Hu monoclonal antibody 16A11; source: Monoclonal Antibody Facility, Eugene, OR).
Description: mRNA binding proteins with a gene-regulatory function at the posttranscriptional level (Marusich et al. 1994). Antibody used recognizes peptide sequence present in various Hu-proteins, including those specifically expressed in neuronal cells (i.e., HuD, HuC; Park et al. 2000). Expression in early differentiating neuronal cells.

Protein Expression Domains in the 3-Day Zebrafish Brain

General appearance: The number of Hu-positive cells has greatly increased compared to 2 days. With the exception of the ventricularly located proliferation zones, the Hu-expression domains now cover most gray matter cells of the entire CNS and most cells in the PNS. Regions with ongoing strong proliferation show less Hu-positive cells with weaker staining compared with more mature regions. However, some clearly postmitotic, differentiated areas remain (as an exception) Hu negative (e.g., retinal outer nuclear layer, subcommissural organ, M2, Mauthner cells, rhombencephalic floor plate cells).

Description of levels on facing page

Sensory organs: Many strongly Hu-positive cells are seen in the olfactory epithelium.

CNS: In the telencephalon, many Hu-stained cells are present in the olfactory bulb (which exhibits glomeruli, i.e., globular neuropil structures formed both by axonal arborizations of olfactory epithelial sensory cells and by dendrites of olfactory bulb neurons), as well as in the gray matter of pallium and (ventral and dorsal divisions) of subpallium. Subpallial and especially pallial Hu domains lie at a distance from the ventricular surface at all anteroposterior levels. Note particularly strongly stained Hu cells in M4, an early migrated cell aggregate possibly representing the future lateral nucleus of the ventral telencephalic area.

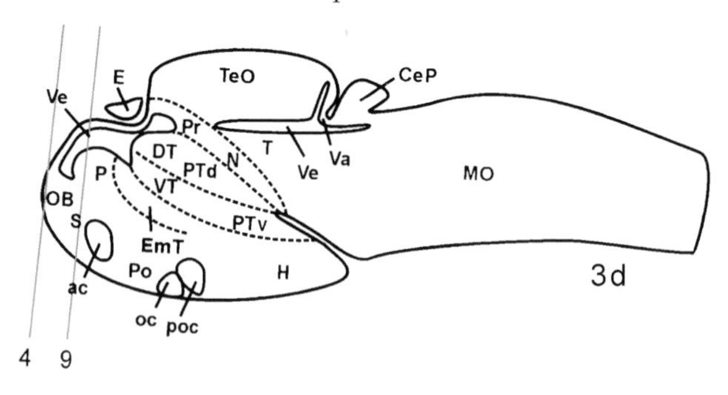

Abbreviations

ac	anterior commissure	pc	posterior commissure
CeP	cerebellar plate	Pi	pigment
DT	dorsal thalamus (thalamus)	Po	preoptic region
E	epiphysis	poc	postoptic commissure
EmT	eminentia thalami	Pr	pretectum
EPi	eye pigment	PTd	dorsal part of posterior tuberculum
Gl	olfactory bulb glomeruli		
H	hypothalamus	PTv	ventral part of posterior tuberculum
Ha	habenula		
lfb	lateral forebrain bundle	S	subpallium
M4	telencephalic migrated area	Sd	dorsal division of S
MO	medulla oblongata	Sv	ventral division of S
N	region of the nucleus of medial longitudinal fascicle	T	midbrain tegmentum
		TeO	tectum opticum
OB	olfactory bulb	TVe	telencephalic ventricle
oc	optic chiasma	Va	valvula cerebelli
OE	olfactory epithelium	Ve	brain ventricle
P	pallium	VT	ventral thalamus (prethalamus)

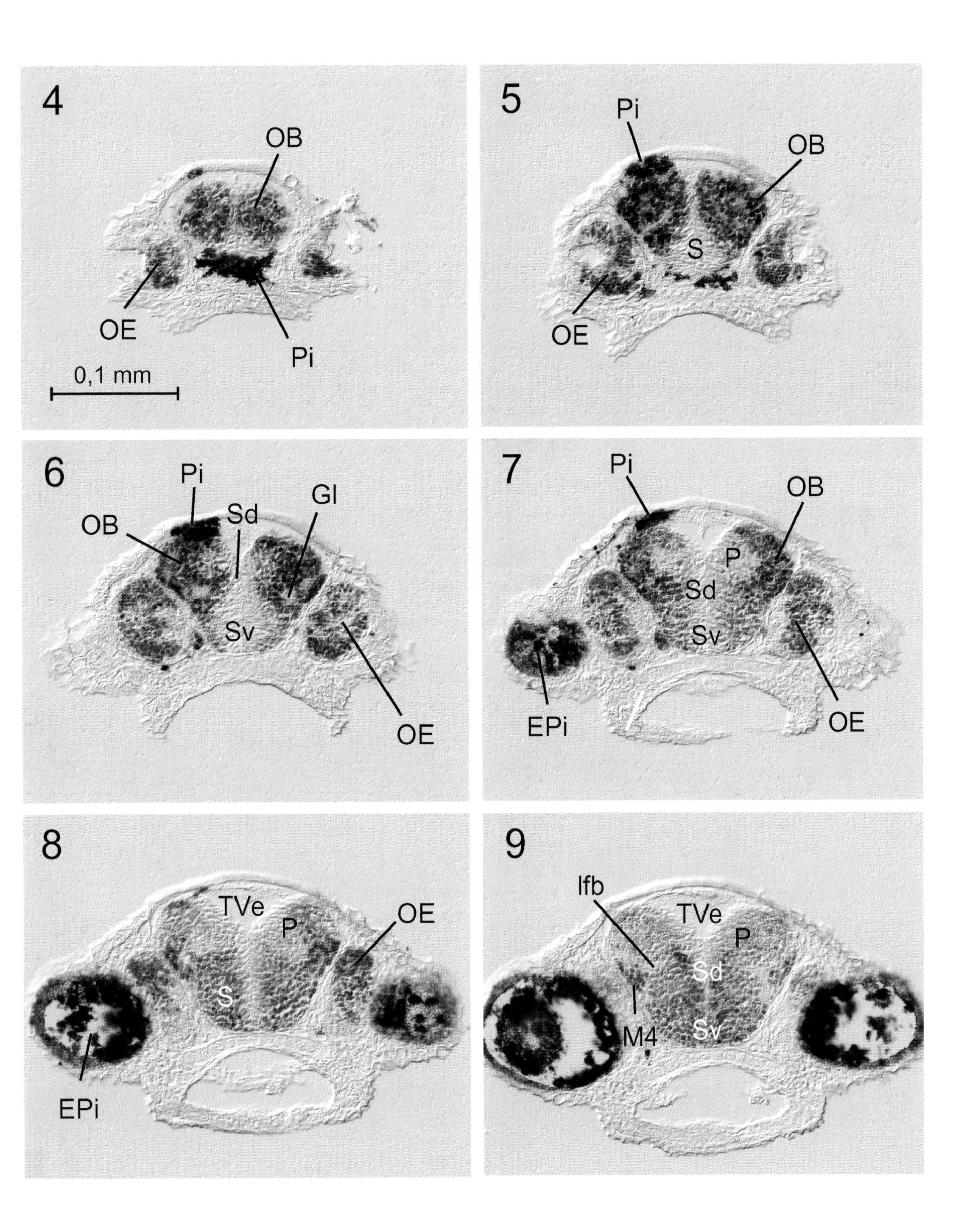
4
OB
OE
Pi
0,1 mm
5
Pi
OB
S
OE
6
Pi
Sd
Gl
OB
Sv
OE
7
Pi
OB
P
Sd
Sv
EPi
OE
8
TVe
P
OE
S
EPi
9
Ifb
TVe
P
Sd
M4
Sv

Hu-proteins (continued, 2nd plate)

Protein Expression Domains in the 3-Day Zebrafish Brain

Description of levels on facing page

Sensory organs: Many Hu-positive cells are seen in ganglionic and inner nuclear layers of retina, but not in outer nuclear layer and retinal peripheral edge.

CNS: Telencephalic Hu-stained cells are present in the gray matter of pallium and ventral and dorsal divisions of subpallium. Note particularly strongly stained Hu cells in M4, an early migrated cell aggregate possibly representing the future lateral nucleus of the ventral telencephalic area. Telencephalic Hu domains (especially pallial ones) lie at a distance from the ventricular surface at all anteroposterior levels. Note that distinct Hu domain of dorsal subpallium continues dorsal to anterior commissure, i.e., at supracommissural levels. Posterior to the anterior commissure, the (strongly proliferative) preoptic region emerges with its rostral pole showing many Hu cells, but these become restricted to the periphery more caudally (as at 2 days). Distinct Hu-positive cell stream visibly arches around the lateral forebrain bundle in the area of the eminentia thalami (future entopeduncular complex, M3). In the diencephalon, Hu-stained cells are present in the habenula and epiphysis, but the (strongly proliferative) dorsal thalamus remains Hu free at these levels, as does the (postmitotic) subcommissural organ.

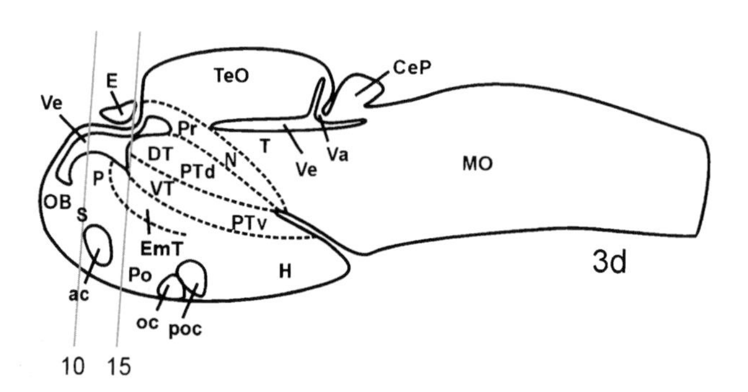

Abbreviations

ac	anterior commissure	Po	preoptic region
CeP	cerebellar plate	poc	postoptic commissure
DT	dorsal thalamus (thalamus)	Pr	pretectum
E	epiphysis	PTd	dorsal part of posterior tuberculum
EmT	eminentia thalami	PTv	ventral part of posterior tuberculum
H	hypothalamus	S	subpallium
Ha	habenula	SCO	subcommissural organ
lfb	lateral forebrain bundle	Sd	dorsal division of S
M3	migrated area of EmT	Sv	ventral division of S
M4	telencephalic migrated area	T	midbrain tegmentum
MO	medulla oblongata	TeO	tectum opticum
N	region of the nucleus of medial longitudinal fascicle	TVe	telencephalic ventricle
OB	olfactory bulb	Va	valvula cerebelli
oc	optic chiasma	Ve	brain ventricle
P	pallium	VT	ventral thalamus (prethalamus)
Pi	pigment		

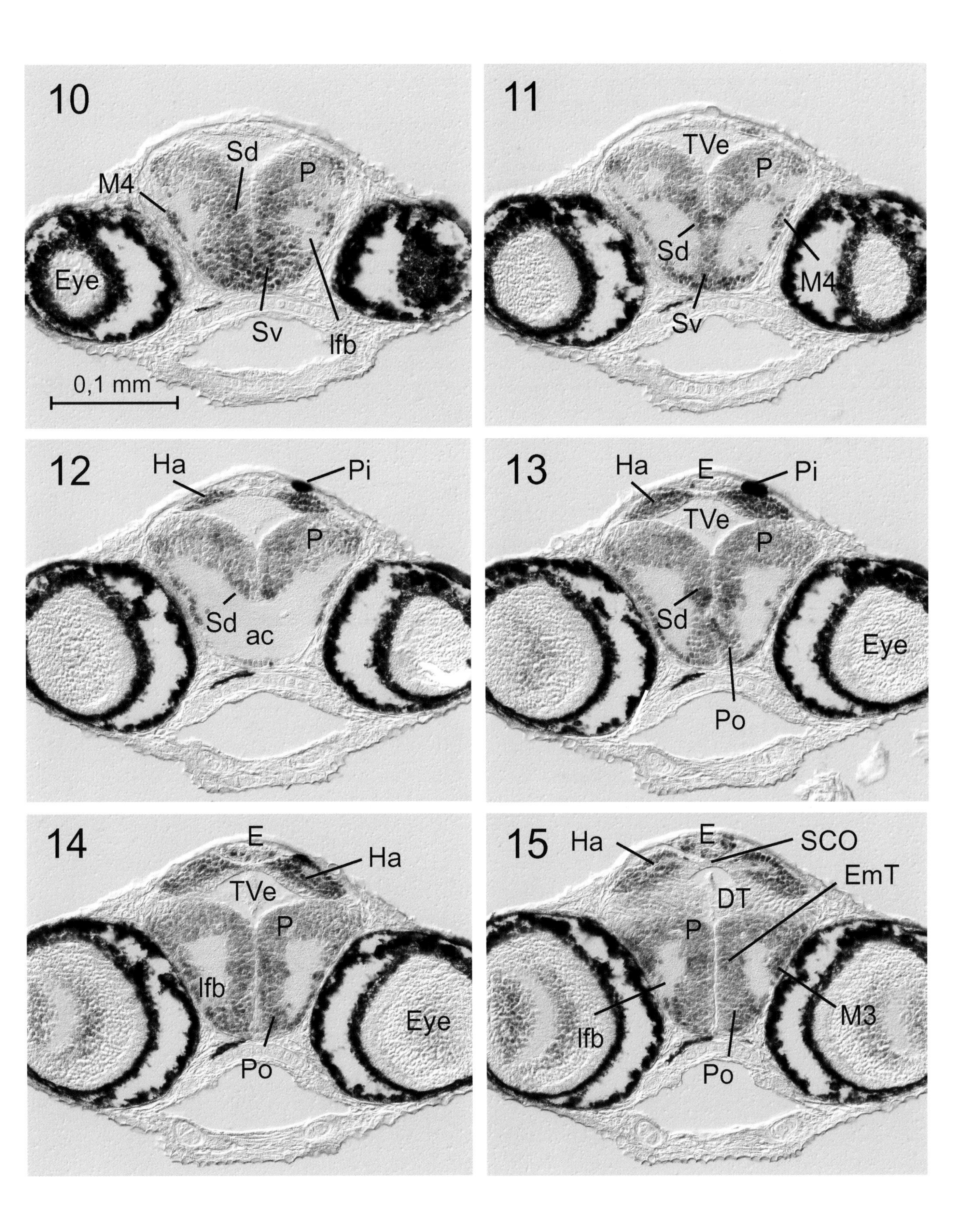
10
Sd
P
M4
Eye
Sv
lfb
0,1 mm
11
TVe
P
Sd
M4
Sv
12
Ha
Pi
P
Sd
ac
13
Ha
E
Pi
TVe
P
Sd
Eye
Po
14
E
Ha
TVe
P
lfb
Eye
Po
15
Ha
E
SCO
EmT
DT
P
M3
lfb
Po

Hu-proteins (continued, 3rd plate)

Protein Expression Domains in the 3-Day Zebrafish Brain

Description of levels on facing page

Sensory organs: Many Hu-positive cells are seen in ganglionic and inner nuclear layers of retina, but not in outer nuclear layer and retinal peripheral edge.

CNS: The (strongly proliferative) preoptic region shows Hu cells at the periphery (as at 2 days), and many more at its caudal pole. Distinct Hu-positive cell stream visibly arches around the lateral forebrain bundle in the area of the eminentia thalami (future entopeduncular complex, M3). In the diencephalon, Hu-stained cells are present in the habenula and epiphysis, but not in the (postmitotic) subcommissural organ. The (strongly proliferative) dorsal and ventral thalami remain Hu free at rostral levels, but increasingly more Hu-positive cells emerge at more caudal thalamic levels. Note the Hu-positive zona limitans intrathalamica, a signaling center between dorsal and ventral thalami, visible at rostral thalamic levels. Also, Hu cells are abundant in the pretectum and its peripherally migrated region (future superficial pretectum, M1). There are few, weakly stained Hu cells in the peripheral rostral optic tectum. The strongly stained Hu cells at the base of the anterior optic tectum possibly represent the griseum tectale, a part of the alar plate mesencephalon curving rostrally below the tectal ventricle.

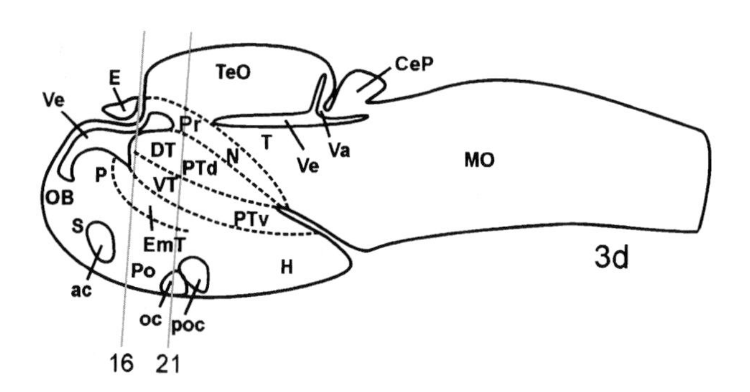

Abbreviations

ac	anterior commissure	P	pallium
CeP	cerebellar plate	pc	posterior commissure
DT	dorsal thalamus (thalamus)	Pi	pigment
fr	fasciculus retroflexus	Po	preoptic region
E	epiphysis	poc	postoptic commissure
EmT	eminentia thalami	Pr	pretectum
GT	griseum tectale	PTd	dorsal part of posterior tuberculum
H	hypothalamus	PTv	ventral part of posterior tuberculum
Ha	habenula	S	subpallium
lfb	lateral forebrain bundle	SCO	subcommissural organ
M1	migrated pretectal area	T	midbrain tegmentum
M3	migrated area of EmT	TeO	tectum opticum
MO	medulla oblongata	Va	valvula cerebelli
N	region of the nucleus of medial longitudinal fascicle	Ve	brain ventricle
OB	olfactory bulb	VT	ventral thalamus (prethalamus)
oc	optic chiasma	ZLI	zona limitans intrathalamica

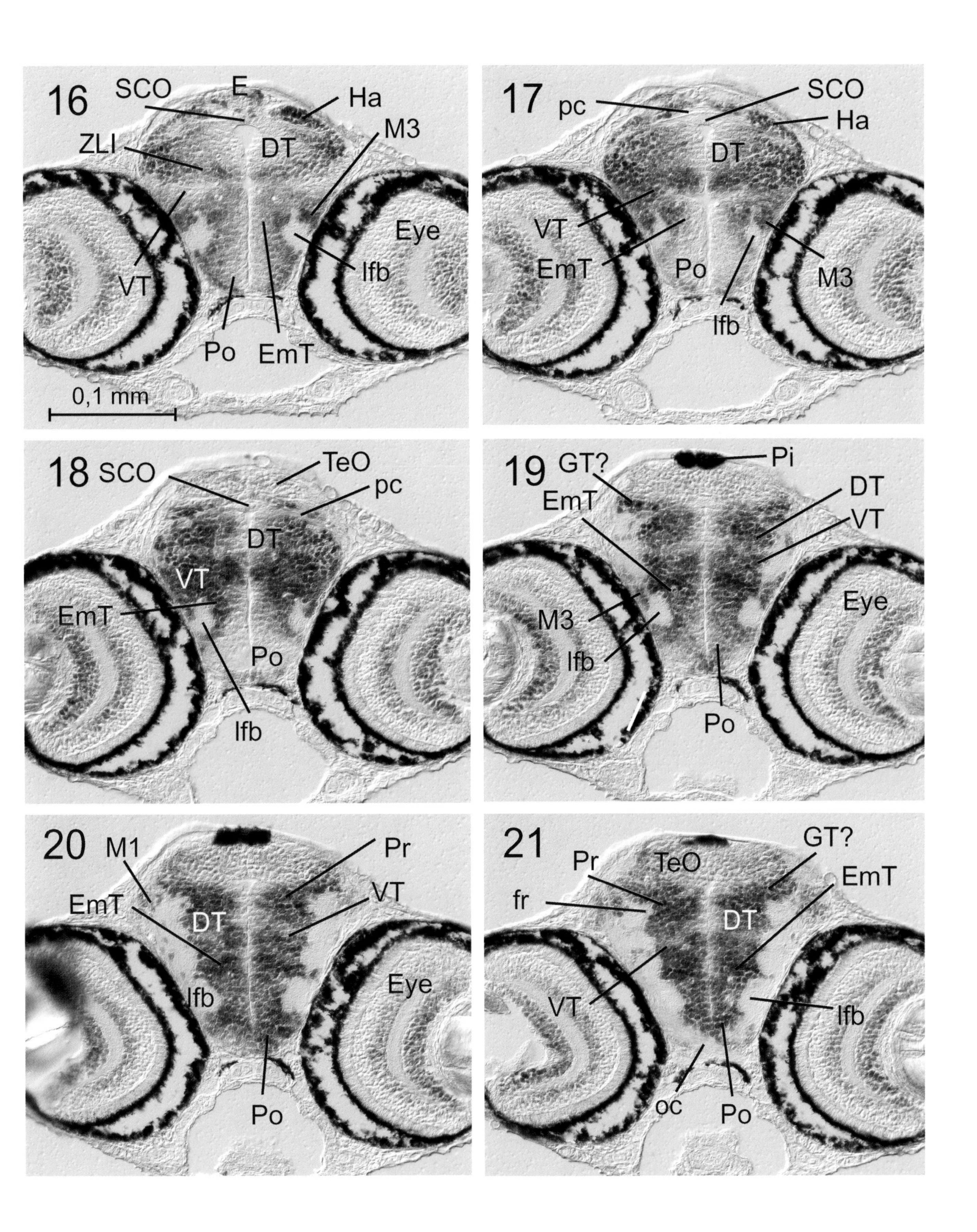
16
SCO
E
Ha
M3
ZLI
DT
Eye
VT
lfb
Po
EmT
0,1 mm
17
pc
SCO
Ha
DT
VT
EmT
Po
M3
lfb
18
SCO
TeO
pc
DT
VT
EmT
Po
lfb
19
GT?
Pi
DT
EmT
VT
Eye
M3
lfb
Po
20
M1
Pr
VT
EmT
DT
Eye
lfb
Po
21
GT?
Pr
TeO
EmT
fr
DT
VT
lfb
oc
Po

Hu-proteins (continued, 4th plate)

Protein Expression Domains in the 3-Day Zebrafish Brain

Description of levels on facing page

Sensory organs: Many Hu-positive cells are seen in ganglionic and inner nuclear layers of retina, but not in outer nuclear layer and retinal peripheral edge.

CNS: At these levels, the preoptic region shows many Hu cells at its caudal pole. In the diencephalon, many Hu-stained cells are present in the pretectum (including its migrated region M1; future superficial pretectum) and dorsal and ventral thalami, as well as in the posterior tuberculum (characterized by its less laterally extending gray matter compared to dorsal thalamus; dorsal and ventral posterior tubercular divisions hard to distinguish). Note peripheral Hu-positive cells in the migrated posterior tubercular region M2 (future preglomerular complex; now separated from the more medial gray matter). Caudal to the postoptic commissure, the rostral hypothalamus emerges, exhibiting many Hu-positive cells. In the optic tectum, few, weakly stained peripheral Hu-positive cells are seen. The strongly stained Hu cells at the base of the anterior optic tectum possibly represent the griseum tectale, a part of the alar plate mesencephalon curving rostrally below the tectal ventricle.

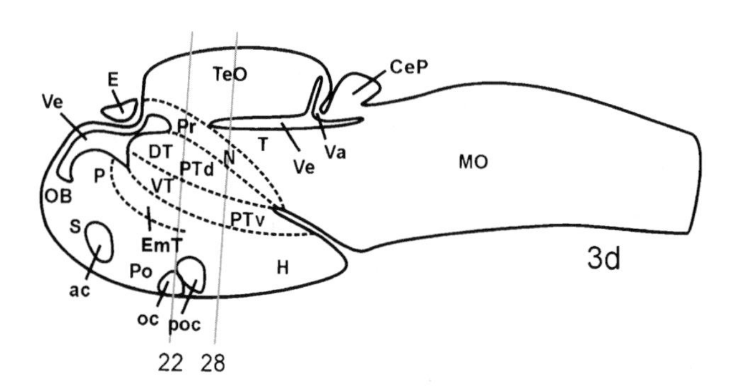

Abbreviations

ac	anterior commissure	oc	optic chiasma
CeP	cerebellar plate	on	optic nerve
DT	dorsal thalamus (thalamus)	P	pallium
E	epiphysis	Pi	pigment
EmT	eminentia thalami	Po	preoptic region
fr	fasciculus retroflexus	poc	postoptic commissure
GT	griseum tectale	Pr	pretectum
H	hypothalamus	PT	posterior tuberculum
Ha	habenula	PTd	dorsal part of posterior tuberculum
Hr	rostral hypothalamus	PTv	ventral part of posterior tuberculum
lfb	lateral forebrain bundle	S	subpallium
M1	migrated pretectal area	T	midbrain tegmentum
M2	migrated posterior tubercular area	TeO	tectum opticum
MO	medulla oblongata	Va	valvula cerebelli
N	region of the nucleus of medial longitudinal fascicle	Ve	brain ventricle
OB	olfactory bulb	VT	ventral thalamus (prethalamus)

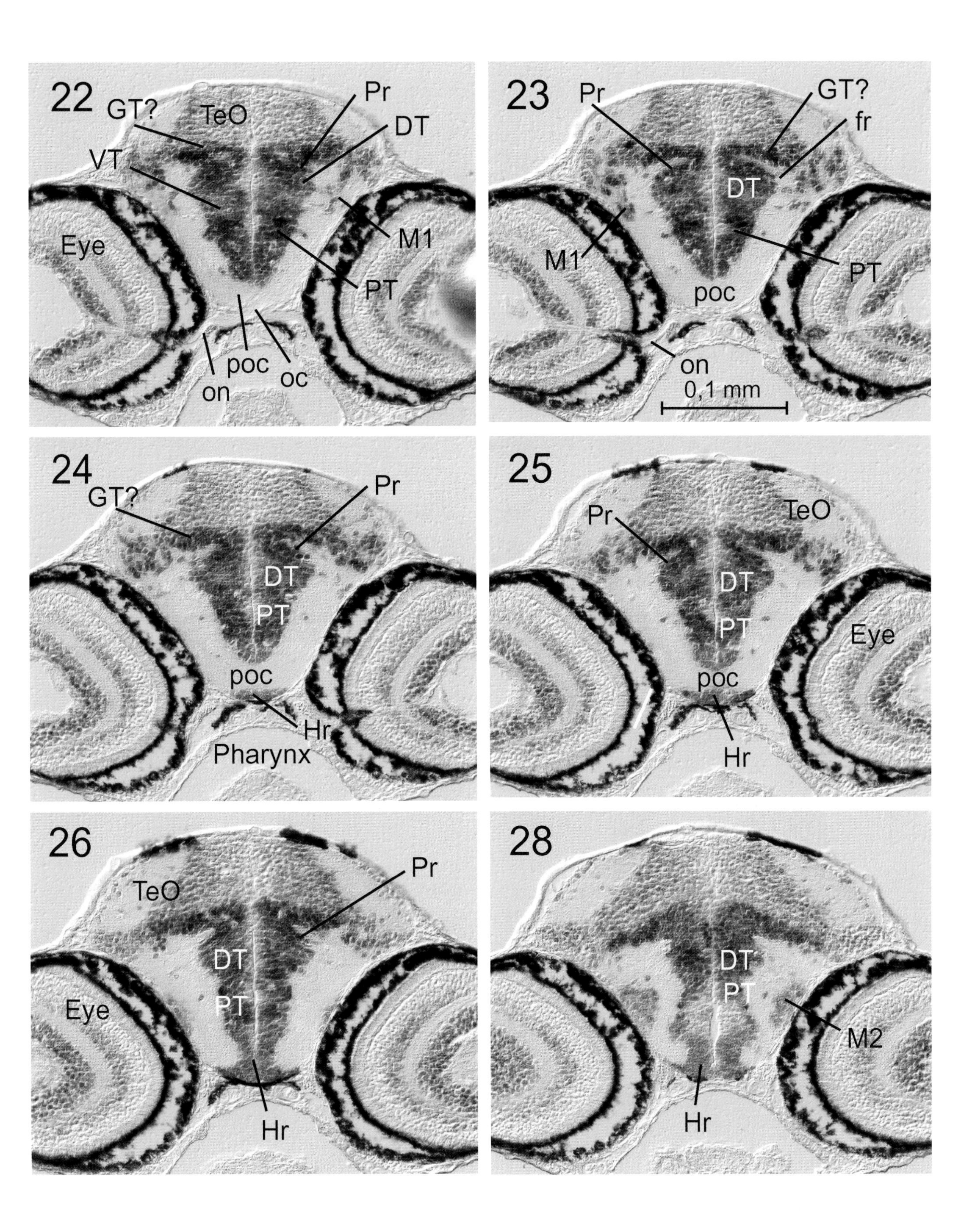
22
GT?
TeO
Pr
DT
VT
M1
Eye
PT
poc
on
oc
23
Pr
GT?
fr
DT
M1
PT
poc
on
0,1 mm
24
GT?
Pr
DT
PT
poc
Hr
Pharynx
25
Pr
TeO
DT
PT
Eye
poc
Hr
26
TeO
Pr
DT
PT
Eye
Hr
28
DT
PT
M2
Hr

Hu-proteins (continued, 5th plate)

Protein Expression Domains in the 3-Day Zebrafish Brain

Description of levels on facing page

Sensory organs: Many Hu-positive cells are seen in ganglionic and inner nuclear layers of retina, but not in outer nuclear layer and retinal peripheral edge.

CNS: At these levels, the diencephalon exhibits Hu-stained cells in the area of the nucleus of the medial longitudinal fascicle (basal synencephalon), in the posterior tuberculum (primarily ventral division visible on facing page, extending into dorsal part of inferior lobe). Note peripheral Hu-positive (as well as many Hu-negative, but postmitotic) cells in the migrated posterior tubercular region M2 (future preglomerular complex; now separated from the more medial gray matter). Caudal to the rostral hypothalamus, the intermediate hypothalamus (characterized by the lateral recess ventricle) emerges, also exhibiting many Hu-positive cells (more anteriorly, less posteriorly). In the optic tectum, few, weakly stained peripheral Hu-positive cells are seen. Caudal to the basal synencephalon, the basal plate midbrain tegmentum appears and contains many strongly stained Hu cells. Note well differentiated (Hu-positive) oculomotor nerve nucleus.

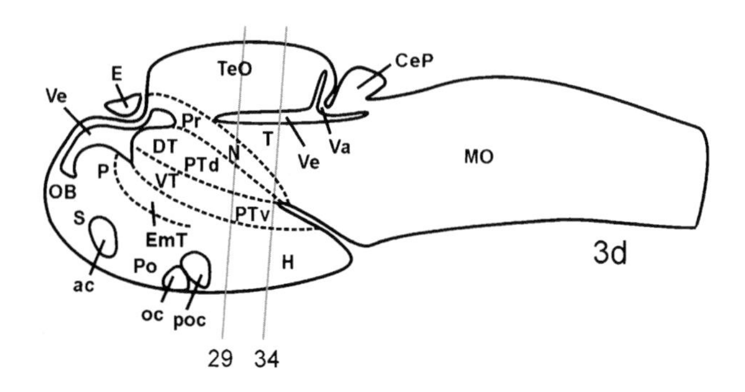

Abbreviations

ac	anterior commissure	OB	olfactory bulb
CeP	cerebellar plate	oc	optic chiasma
DT	dorsal thalamus (thalamus)	P	pallium
E	epiphysis	Po	preoptic region
EmT	eminentia thalami	poc	postoptic commissure
H	hypothalamus	Pr	pretectum
Hi	intermediate hypothalamus	PTd	dorsal part of posterior tuberculum
Hr	rostral hypothalamus	PTv	ventral part of posterior tuberculum
LVe	lateral recess ventricle of H	S	subpallium
M2	migrated posterior tubercular area	T	midbrain tegmentum
mlf	medial longitudinal fascicle	TeO	tectum opticum
MO	medulla oblongata	Va	valvula cerebelli
N	region of the nucleus of medial longitudinal fascicle	Ve	brain ventricle
NIII	oculomotor nerve nucleus	VT	ventral thalamus (prethalamus)

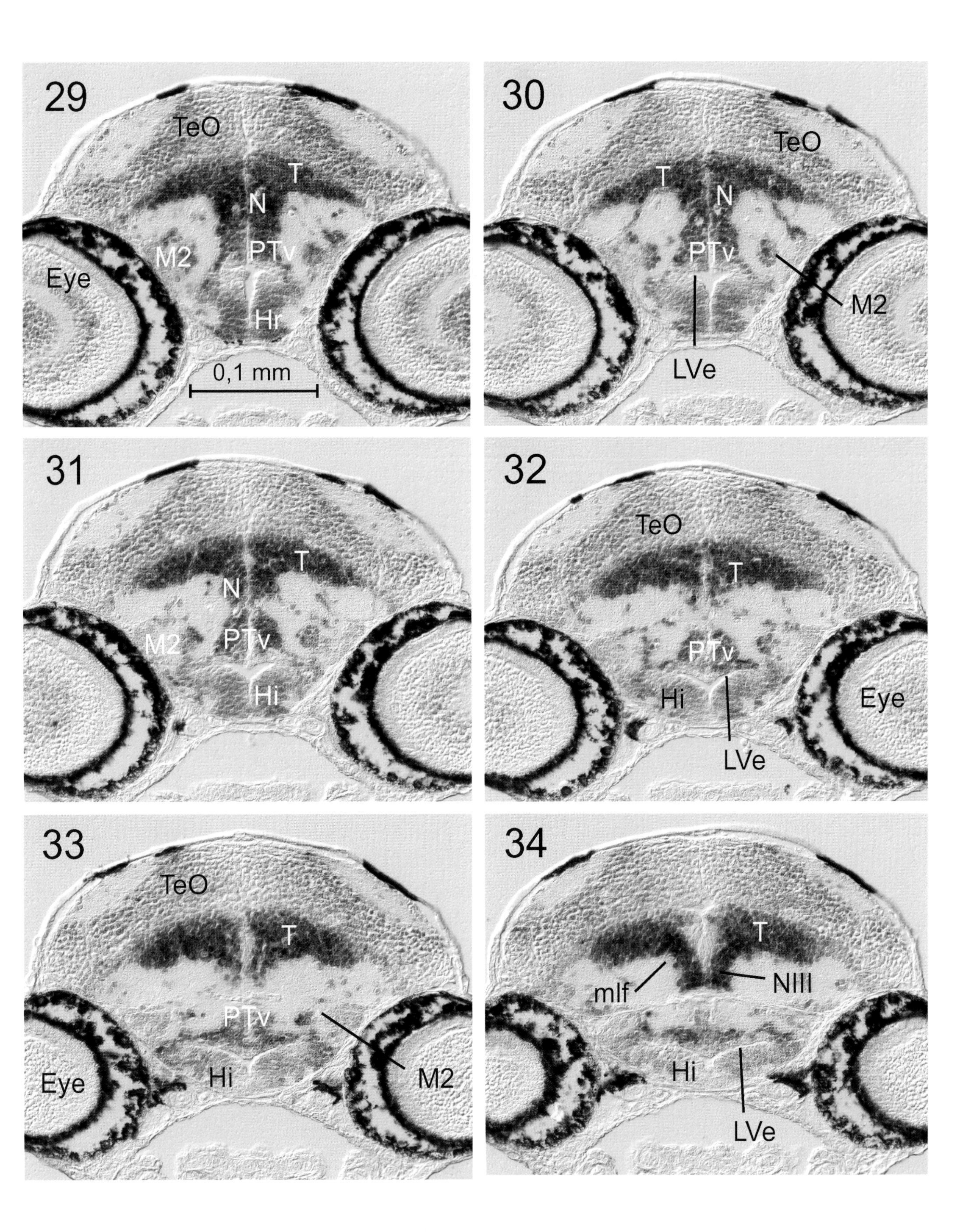
29
TeO
T
N
M2
PTv
Eye
Hr
0,1 mm
30
TeO
T
N
PTv
M2
LVe
31
T
N
M2
PTv
Hi
32
TeO
T
PTv
Hi
Eye
LVe
33
TeO
T
PTv
M2
Eye
Hi
34
T
mlf
NIII
Hi
LVe

Hu-proteins (continued, 6th plate)

Protein Expression Domains in the 3-Day Zebrafish Brain

Description of levels on facing page

Sensory organs/PNS: Many darkly stained Hu cells are present in various cranial nerve ganglia.

CNS: At these levels, the caudal hypothalamus (characterized by the posterior recess ventricle) replaces the ventral part of the posterior tuberculum, and continues into the most caudal part of the inferior lobe where it also replaces the intermediate hypothalamus. The caudal hypothalamus exhibits many Hu-positive cells (especially anteriorly, much less posteriorly). In the optic tectum, few, weakly stained peripheral Hu-positive cells are seen, but they increasingly disappear in the more caudal sections. The torus semicircularis contains many weakly to strongly stained Hu neurons. Caudal to midbrain tegmentum, the region of the (brainstem) midbrain–hindbrain boundary appears, containing many strongly stained Hu cells, as does the caudally adjacent medulla oblongata. Also the emerging valvula contains weakly stained Hu cells. Note well-differentiated oculomotor nerve nucleus and interpeduncular nucleus (both Hu positive) and Hu-negative floor plate cells at the bottom of rhombencephalic ventricle. The hypophysis is free of Hu signal.

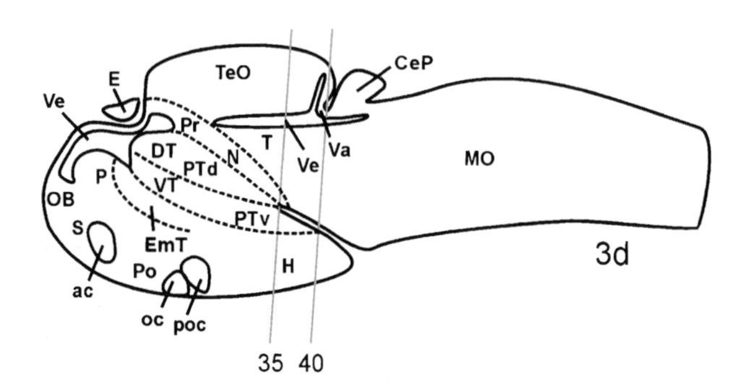

Abbreviations

ac	anterior commissure	NIn	nucleus interpeduncularis
ALLG	anterior lateral line ganglion	OB	olfactory bulb
cec	cerebellar commissure	oc	optic chiasma
CeP	cerebellar plate	P	pallium
Ch	chorda dorsalis	Po	preoptic region
DT	dorsal thalamus (thalamus)	poc	postoptic commissure
E	epiphysis	Pr	pretectum
EmT	eminentia thalami	PTd	dorsal part of posterior tuberculum
EPi	eye pigment	PTv	ventral part of posterior tuberculum
FP	floor plate		
H	hypothalamus	PVe	posterior recess ventricle of H
Hc	caudal hypothalamus	RVe	rhombencephalic ventricle
Hi	intermediate hypothalamus	S	subpallium
Hy	hypophysis (pituitary)	T	midbrain tegmentum
LVe	lateral recess ventricle of H	TeO	tectum opticum
mlf	medial longitudinal fascicle	TG	trigeminal ganglion
MO	medulla oblongata	TS	torus semicircularis
N	region of the nucleus of medial longitudinal fascicle	Va	valvula cerebelli
		Ve	brain ventricle
NIII	oculomotor nerve nucleus	VT	ventral thalamus (prethalamus)

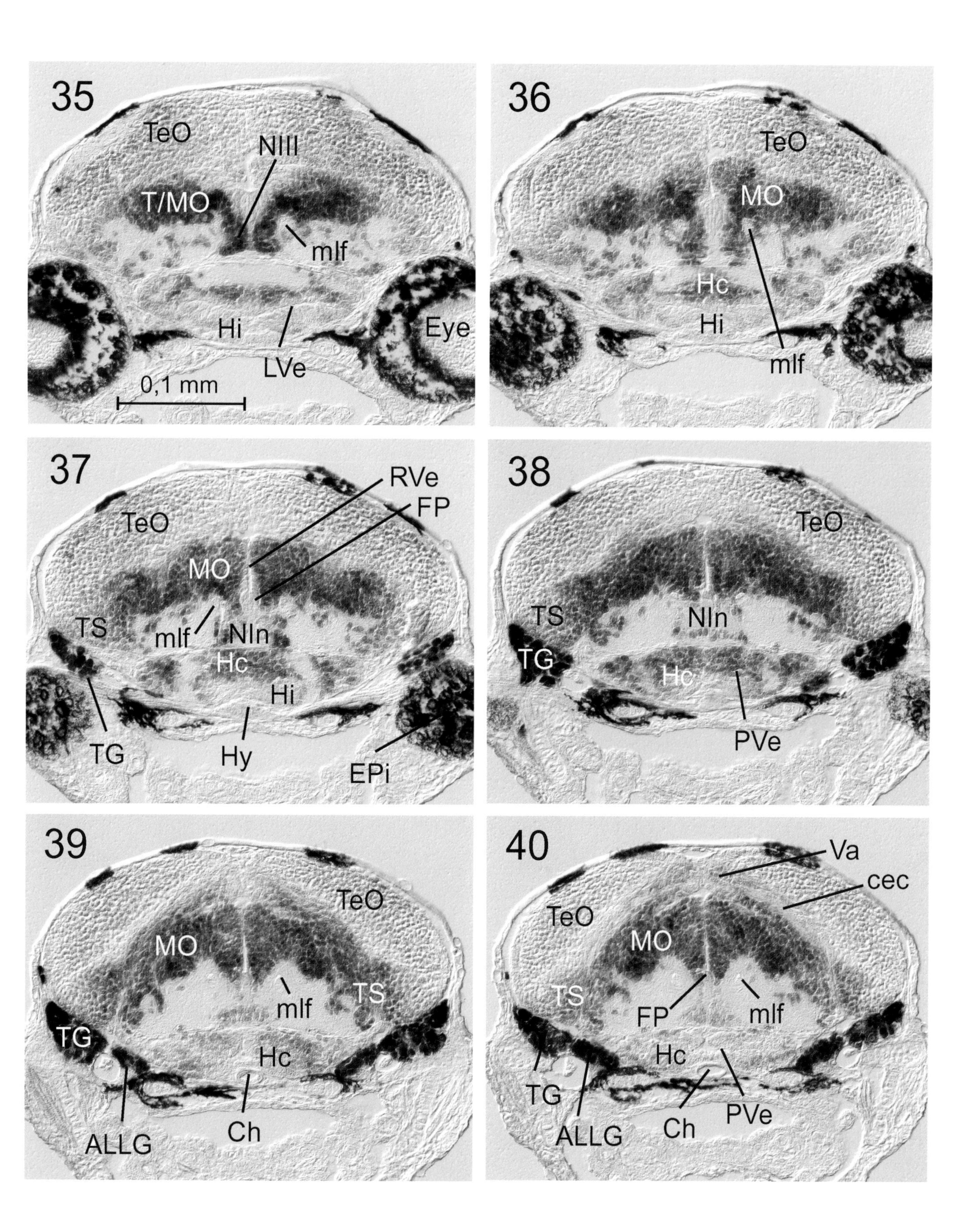
35
TeO
NIII
T/MO
mlf
Hi
LVe
Eye
0,1 mm
36
TeO
MO
Hc
Hi
mlf
37
RVe
FP
TeO
MO
TS
mlf
NIn
Hc
Hi
TG
Hy
EPi
38
TeO
TS
NIn
TG
Hc
PVe
39
TeO
MO
mlf
TS
TG
Hc
ALLG
Ch
40
Va
cec
TeO
MO
TS
FP
mlf
Hc
TG
ALLG
Ch
PVe

Hu-proteins (continued, 7th plate)

Protein Expression Domains in the 3-Day Zebrafish Brain

Description of levels on facing page

Sensory organs/PNS: Many darkly stained Hu cells are present in various cranial nerve ganglia.

CNS: At these levels, the caudal hypothalamus (characterized by the posterior recess ventricle) fades out, showing only very few, faintly stained, peripheral Hu cells. The (strongly proliferative) caudal optic tectum fails to show Hu-positive cells. The valvula cerebelli, cerebellar plate and especially the eminentia granularis contain many weakly stained Hu-positive cells (but increasingly fewer Hu cells in the caudal cerebellar plate). The medulla oblongata contains many strongly stained Hu cells, as do some identifiably differentiated medullary structures, namely the superior raphe and superior reticular formation. Note that the differentiated Mauthner neuron is Hu free.

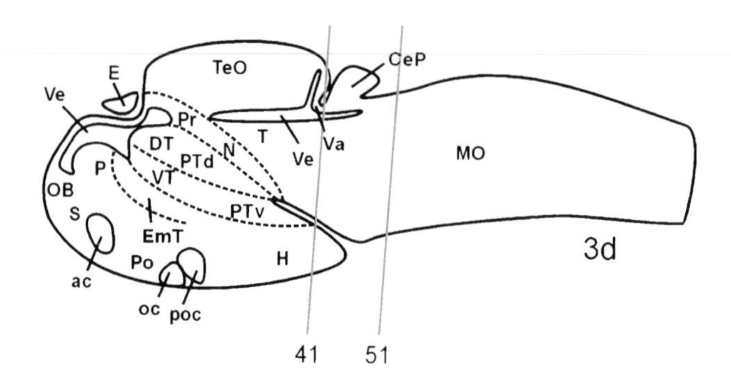

Abbreviations

ac	anterior commissure	oc	optic chiasma
ALLG	anterior lateral line ganglion	OC	otic capsule
CC	cerebellar crest	OG	otic ganglion
cec	cerebellar commissure	P	pallium
CeP	cerebellar plate	Pi	pigment
Ch	chorda dorsalis	Po	preoptic region
DT	dorsal thalamus (thalamus)	poc	postoptic commissure
E	epiphysis	Pr	pretectum
EG	eminentia granularis	PTd	dorsal part of posterior tuberculum
EmT	eminentia thalami	PTv	ventral part of posterior tuberculum
FG	facial ganglion	PVe	posterior recess ventricle of H
GG	glossopharyngeal ganglion	RVe	rhombencephalic ventricle
H	hypothalamus	S	subpallium
Hc	caudal hypothalamus	SR	superior raphe
LHP	lateral hinge-point between cerebellum/medulla oblongata	SRF	superior reticular formation
mlf	medial longitudinal fascicle	T	midbrain tegmentum
MN	Mauthner neuron	TeO	tectum opticum
MO	medulla oblongata	TG	trigeminal ganglion
N	region of the nucleus of medial longitudinal fascicle	Va	valvula cerebelli
OB	olfactory bulb	Ve	brain ventricle
		VT	ventral thalamus (prethalamus)

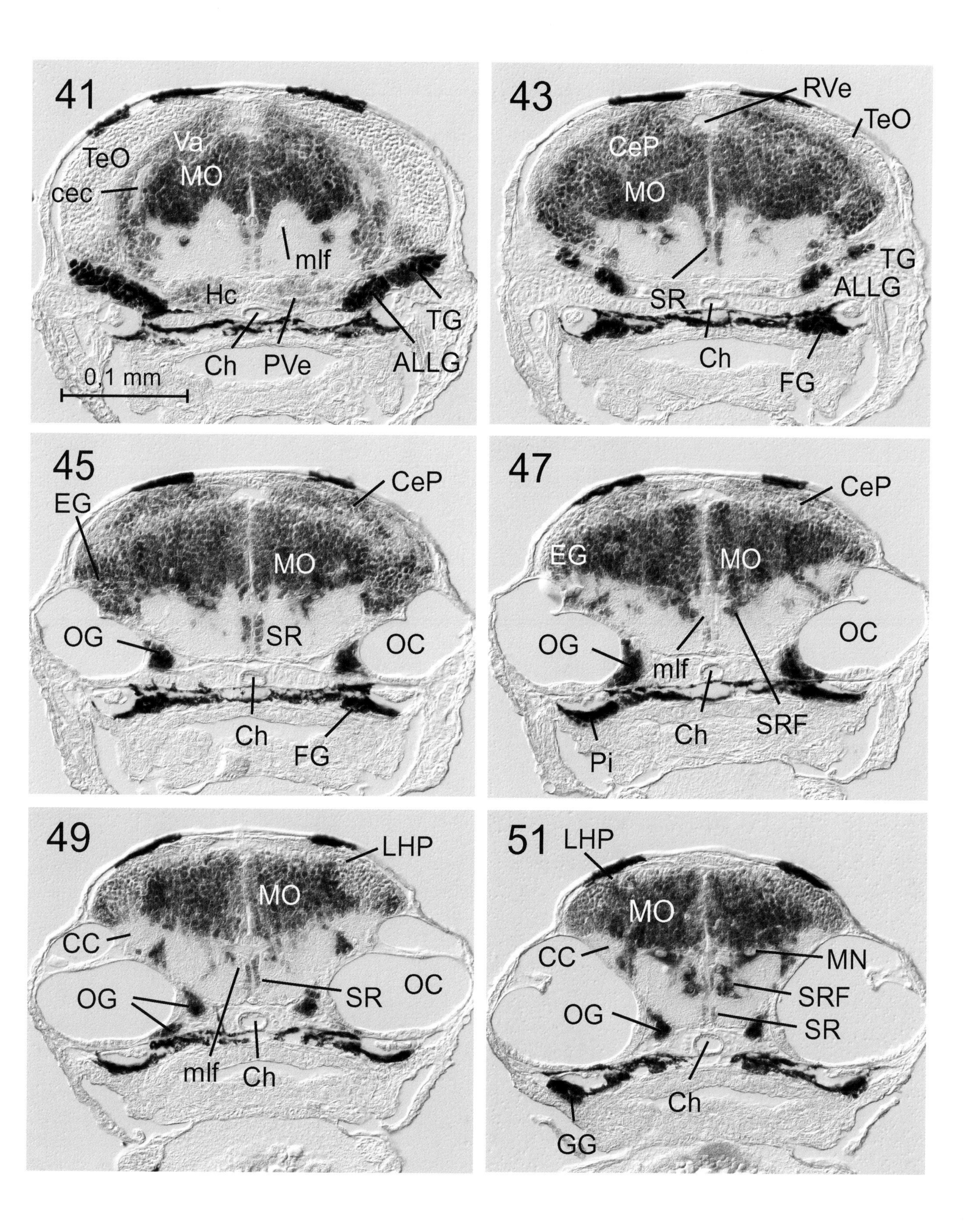
41
Va
TeO
MO
cec
mlf
Hc
TG
Ch
PVe
ALLG
0,1 mm
43
RVe
TeO
CeP
MO
TG
ALLG
SR
Ch
FG
45
CeP
EG
MO
OG
SR
OC
Ch
FG
47
CeP
EG
MO
OC
OG
mlf
Ch
SRF
Pi
49
LHP
MO
CC
OC
OG
SR
mlf
Ch
51
LHP
MO
CC
MN
SRF
OG
SR
Ch
GG

Hu-proteins (continued, 8th plate)

Protein Expression Domains in the 3-Day Zebrafish Brain

Description of levels on facing page

Sensory organs/PNS: Many darkly stained Hu cells are present in various cranial nerve ganglia.

CNS: At otic levels, the medulla oblongata continues to exhibit many strongly stained Hu cells, but the (proliferative) rhombic lip remains Hu free, as also many cells in the lateral medulla, appearing to derive from the rhombic lip, but the most ventrolateral of these cells shows an increasingly stronger Hu signal. Some identifiably differentiated medullary structures, namely the intermediate raphe and intermediate reticular formation are strongly Hu positive.

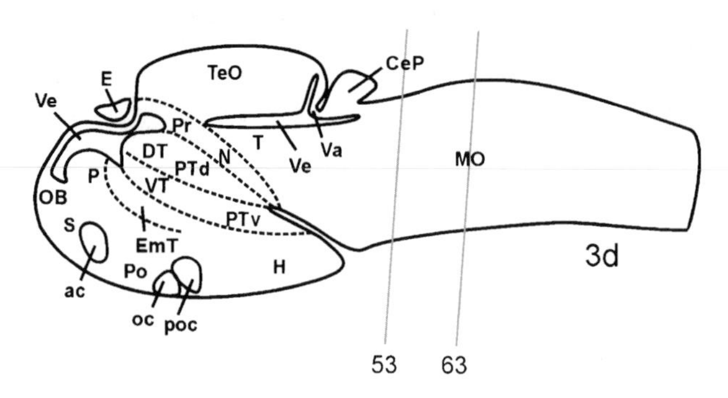

Abbreviations

ac	anterior commissure	P	pallium
CeP	cerebellar plate	Pi	pigment
Ch	chorda dorsalis	PLLG	posterior lateral line ganglion
DT	dorsal thalamus (thalamus)	Po	preoptic region
E	epiphysis	poc	postoptic commissure
EmT	eminentia thalami	Pr	pretectum
GG	glossopharyngeal ganglion	PTd	dorsal part of posterior tuberculum
H	hypothalamus	PTv	ventral part of posterior tuberculum
IMR	Intermediate raphe	RL	rhombic lip
IMRF	intermediate reticular formation	S	subpallium
mlf	medial longitudinal fascicle	T	midbrain tegmentum
MO	medulla oblongata	TeO	tectum opticum
N	region of the nucleus of medial longitudinal fascicle	Va	valvula cerebelli
OB	olfactory bulb	Ve	brain ventricle
oc	optic chiasma	VG	vagal ganglion
OC	otic capsule	VT	ventral thalamus (prethalamus)
OG	otic ganglion		

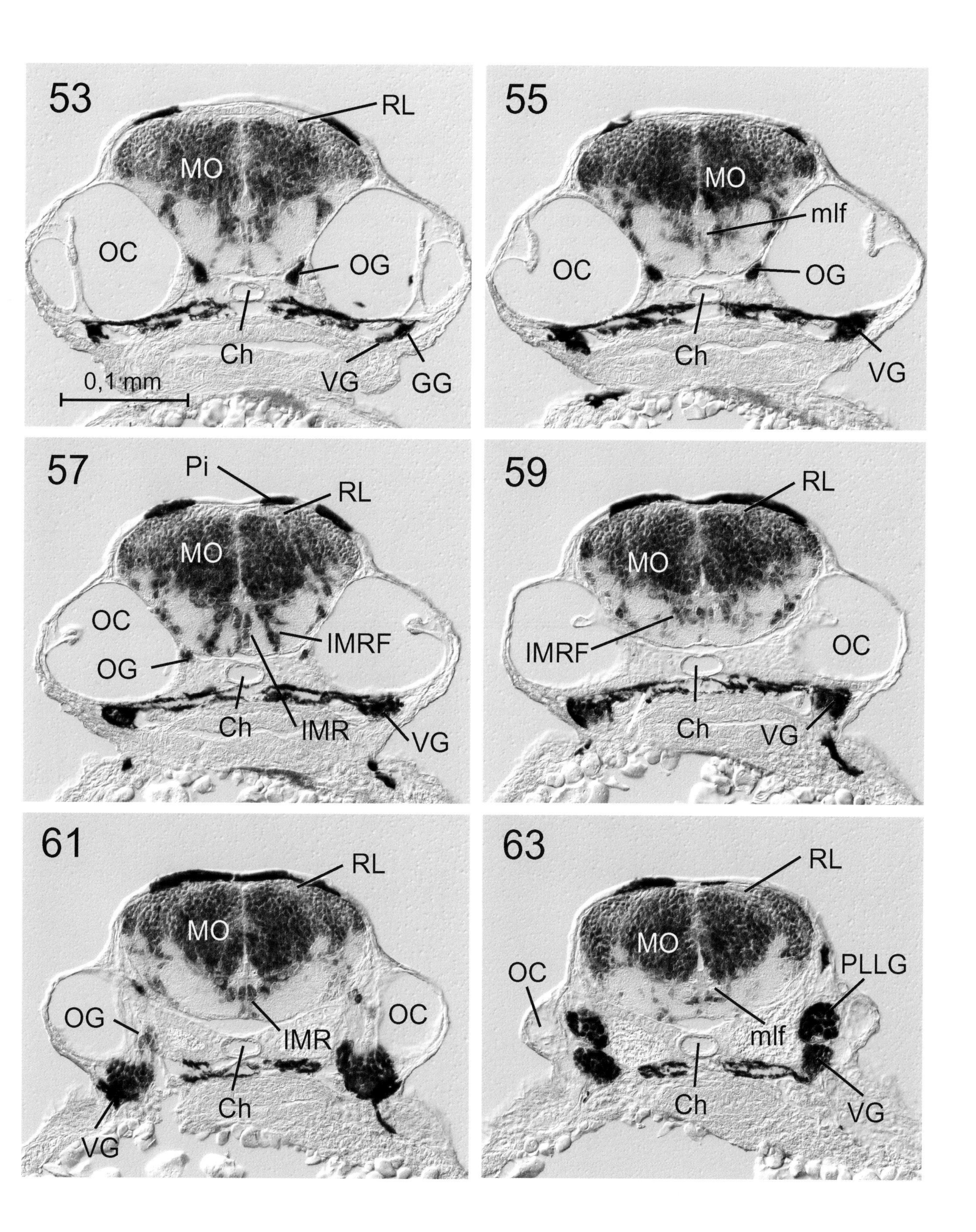
53
RL
MO
OC
OG
Ch
0,1 mm
VG
GG
55
MO
mlf
OC
OG
Ch
VG
57
Pi
RL
MO
OC
IMRF
OG
Ch
IMR
VG
59
RL
MO
IMRF
OC
Ch
VG
61
RL
MO
OG
OC
IMR
Ch
VG
63
RL
MO
OC
PLLG
mlf
Ch
VG

Hu-proteins (continued, 9th plate)

Protein Expression Domains in the 3-Day Zebrafish Brain

Description of levels on facing page

Sensory organs/PNS: Many darkly stained Hu cells are present in various cranial nerve ganglia.

CNS: At postotic levels, the medulla oblongata still contains many strongly stained Hu cells, but the (proliferative) rhombic lip is Hu free, as are many cells in the lateral medulla, appearing to derive from the rhombic lip, but the most ventrolateral of these cells shows an increasingly stronger Hu signal. Some identifiably differentiated medullary structures, namely the inferior olive, inferior raphe and inferior reticular formation are strongly Hu positive.

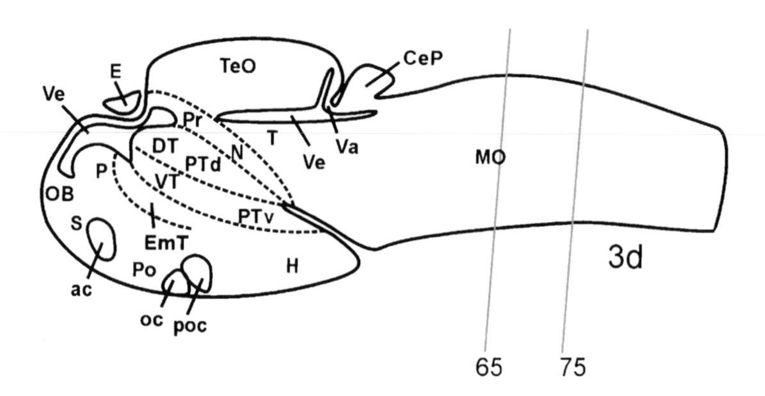

Abbreviations

ac	anterior commissure
CeP	cerebellar plate
Ch	chorda dorsalis
DT	dorsal thalamus (thalamus)
E	epiphysis
EmT	eminentia thalami
H	hypothalamus
IO	inferior olive
IR	inferior raphe
IRF	inferior reticular formation
mlf	medial longitudinal fascicle
MO	medulla oblongata
N	region of the nucleus of medial longitudinal fascicle
OB	olfactory bulb
oc	optic chiasma
P	pallium
Pi	pigment
PLLG	posterior lateral line ganglion
Po	preoptic region
poc	postoptic commissure
Pr	pretectum
PTd	dorsal part of posterior tuberculum
PTv	ventral part of posterior tuberculum
RL	rhombic lip
S	subpallium
T	midbrain tegmentum
TeO	tectum opticum
Va	valvula cerebelli
Ve	brain ventricle
VT	ventral thalamus (prethalamus)

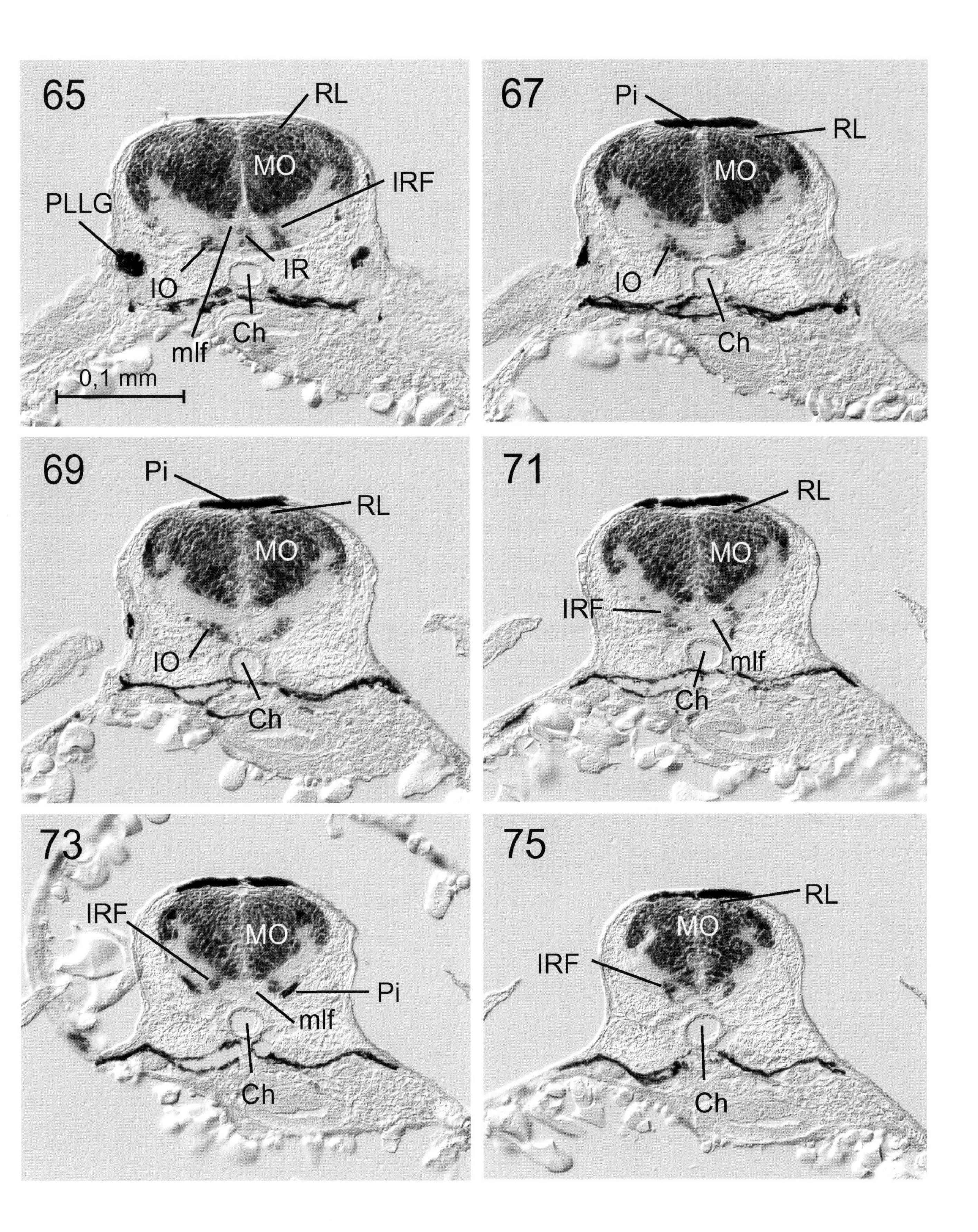
65
RL
MO
IRF
PLLG
IO
IR
Ch
mlf
0,1 mm
67
Pi
RL
MO
IO
Ch
69
Pi
RL
MO
IO
Ch
71
RL
MO
IRF
mlf
Ch
73
IRF
MO
Pi
mlf
Ch
75
RL
MO
IRF
Ch

BrdU (shown with monoclonal mouse antibody against BrdU [5-bromo-2′-deoxyuridine]; source: Dako Diagnostika, Hamburg, Germany).
Description: BrdU is a thymidine analogue incorporated into the DNA during S-phase mitosis, used to visualize proliferating cells.

BrdU Cells in the 5-Day Zebrafish Brain

General appearance: BrdU, if used in a saturation label (explained in the following chapter), reveals all central nervous proliferation zones (essentially corroborating results gained with PCNA immunohistochemistry; see Chapter 3), which typically lie directly at the ventricle along the entire neuraxis, generally medial to cells expressing neuronal determination (e.g., *neurod*) or differentiation (e.g., Hu) markers.

Description of levels on facing page

Sensory organs: BrdU signal is restricted to retinal peripheral edge (note also non-neural BrdU-positive cells in lens). Clusters of BrdU-positive cells are seen in olfactory epithelium.

CNS: In the telencephalon, pallial and subpallial (dorsal and ventral divisions visible on facing page) BrdU cells, including a large BrdU-positive zone in medial proliferative zone of olfactory bulb, lie directly at the ventricular surface at all anteroposterior levels. The anterior commissure separates the subpallium posteriorly from the preoptic region.

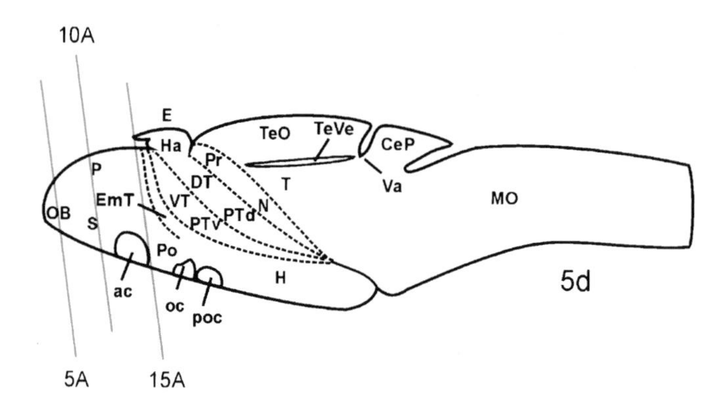

Abbreviations

ac	anterior commissure	Po	preoptic region
CeP	cerebellar plate	poc	postoptic commissure
DT	dorsal thalamus (thalamus)	Pr	pretectum
E	epiphysis	PTd	dorsal part of posterior tuberculum
EmT	eminentia thalami	PTv	ventral part of posterior tuberculum
EPi	eye pigment	S	subpallium
H	hypothalamus	Sd	dorsal division of S
Ha	habenula	Sv	ventral division of S
MO	medulla oblongata	T	midbrain tegmentum
N	region of the nucleus of the medial longitudinal fascicle	TeO	tectum opticum
OB	olfactory bulb	TeVe	tectal ventricle
oc	optic chiasma	TVe	telencephalic ventricle
OE	olfactory epithelium	Va	valvula cerebelli
P	pallium	VT	ventral thalamus (prethalamus)

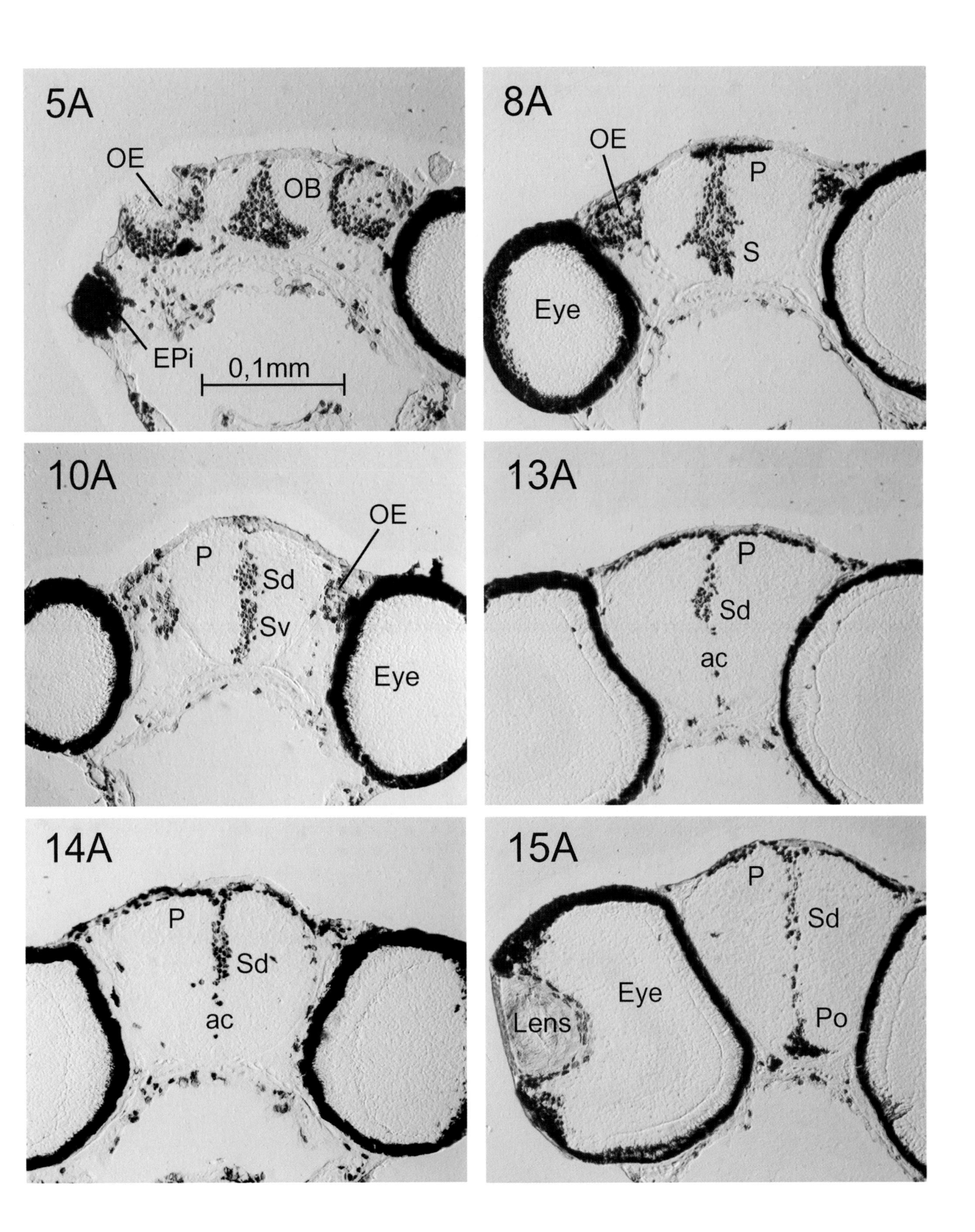
5A
OE
OB
EPi
0,1mm
8A
OE
P
S
Eye
10A
OE
P
Sd
Sv
Eye
13A
P
Sd
ac
14A
P
Sd
ac
15A
P
Sd
Eye
Lens
Po

BrdU (continued, 2nd plate)

BrdU Cells in the 5-Day Zebrafish Brain

Description of levels on facing page

CNS: At these levels, many BrdU-labeled cells are present directly at the ventricular lining of the preoptic region, which is ventrally adjacent to the much thinner BrdU-positive area of the eminentia thalami. The most caudal BrdU cells of the pallium are seen which are replaced caudally by diencephalic, proliferative cell populations at the ventricular lining, i.e., BrdU-positive cells in the basal part of the habenula, the dorsal thalamus proper, and the ventral thalamus (these BrdU cells extend particularly far laterally), but not in the epiphysis. The most anterior optic tectum emerges with its medial BrdU zone; the presumptive torus lateralis appears free of BrdU-labeled cells.

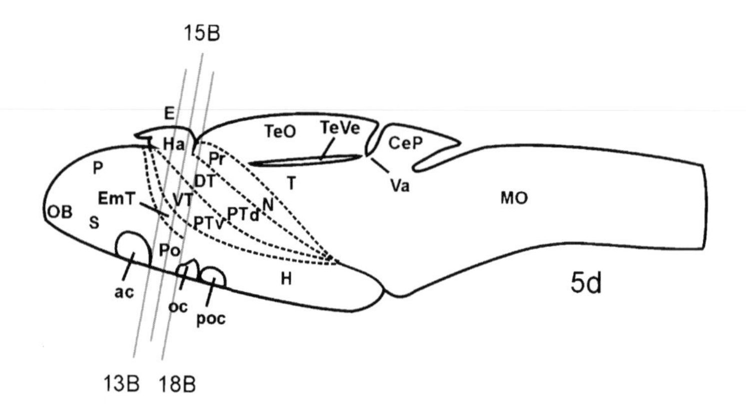

Abbreviations

ac	anterior commissure	Pi	pigment
CeP	cerebellar plate	Po	preoptic region
DT	dorsal thalamus (thalamus)	poc	postoptic commissure
E	epiphysis	Pr	pretectum
EmT	eminentia thalami	PTd	dorsal part of posterior tuberculum
H	hypothalamus		
Ha	habenula	PTv	ventral part of posterior tuberculum
lfb	lateral forebrain bundle		
m	medial tectal proliferation zone	S	subpallium
MO	medulla oblongata	T	midbrain tegmentum
N	region of the nucleus of medial longitudinal fascicle	TeO	tectum opticum
		TeVe	tectal ventricle
OB	olfactory bulb	TL	torus longitudinalis
oc	optic chiasma	Va	valvula
on	optic nerve	VT	ventral thalamus (prethalamus)
P	pallium	ZLI	zona limitans intrathalamica

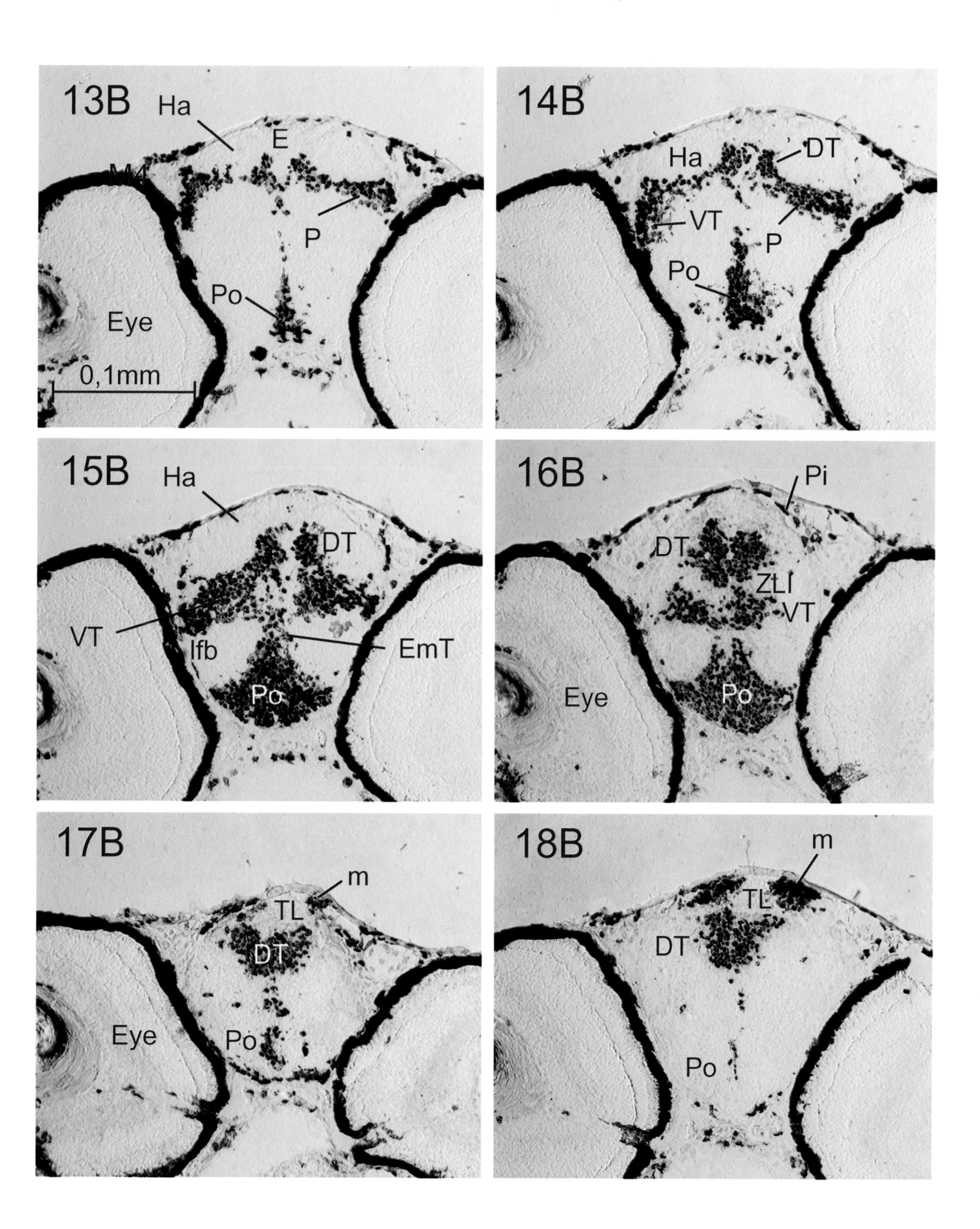
13B
Ha
E
P
Po
Eye
0,1mm
14B
Ha
DT
VT
P
Po
15B
Ha
DT
VT
lfb
EmT
Po
16B
Pi
DT
ZLI
VT
Eye
Po
17B
m
TL
DT
Eye
Po
18B
m
TL
DT
Po

BrdU (continued, 3rd plate)

BrdU Cells in the 5-Day Zebrafish Brain

Description of levels on facing page

CNS: At these levels, the preoptic region becomes separated caudally by the postoptic commissure from the emerging rostral hypothalamus which exhibits many BrdU cells in its ventricular proliferation zone. In the remainder of the diencephalon, such ventricular BrdU-positive zones are present in the dorsal thalamus proper and pretectum, as well as in the posterior tuberculum (dorsal and ventral divisions visible on facing page; interpreted as basal plate of dorsal and ventral thalamic prosomeres/P2 and P3). Within the optic tectum, medial (dorsal and ventral parts visible rostrally) and lateral BrdU domains are present. Note the cluster of BrdU-positive cells in the far migrated posterior tubercular region M2 (future preglomerular complex).

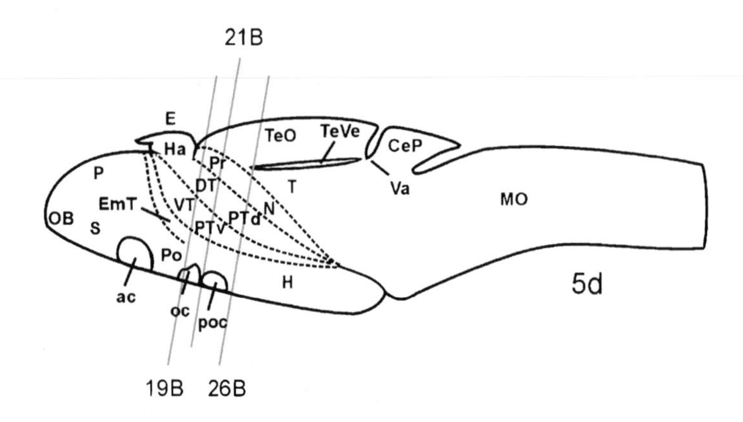

Abbreviations

ac	anterior commissure	OB	olfactory bulb
CeP	cerebellar plate	oc	optic chiasma
DT	dorsal thalamus (thalamus)	on	optic nerve
E	epiphysis	P	pallium
EmT	eminentia thalami	pc	posterior commissure
H	hypothalamus	Po	preoptic region
Ha	habenula	poc	postoptic commissure
Hr	rostral hypothalamus	Pr	pretectum
l	lateral tectal proliferation zone	PTd	dorsal part of posterior tuberculum
M2	migrated posterior tubercular area	PTv	ventral part of posterior tuberculum
m	medial tectal proliferation zone	S	subpallium
md	dorsal part of m	T	midbrain tegmentum
mv	ventral part of m	TeO	tectum opticum
MO	medulla oblongata	TeVe	tectal ventricle
N	region of the nucleus of medial longitudinal fascicle	Va	valvula
		VT	ventral thalamus (prethalamus)

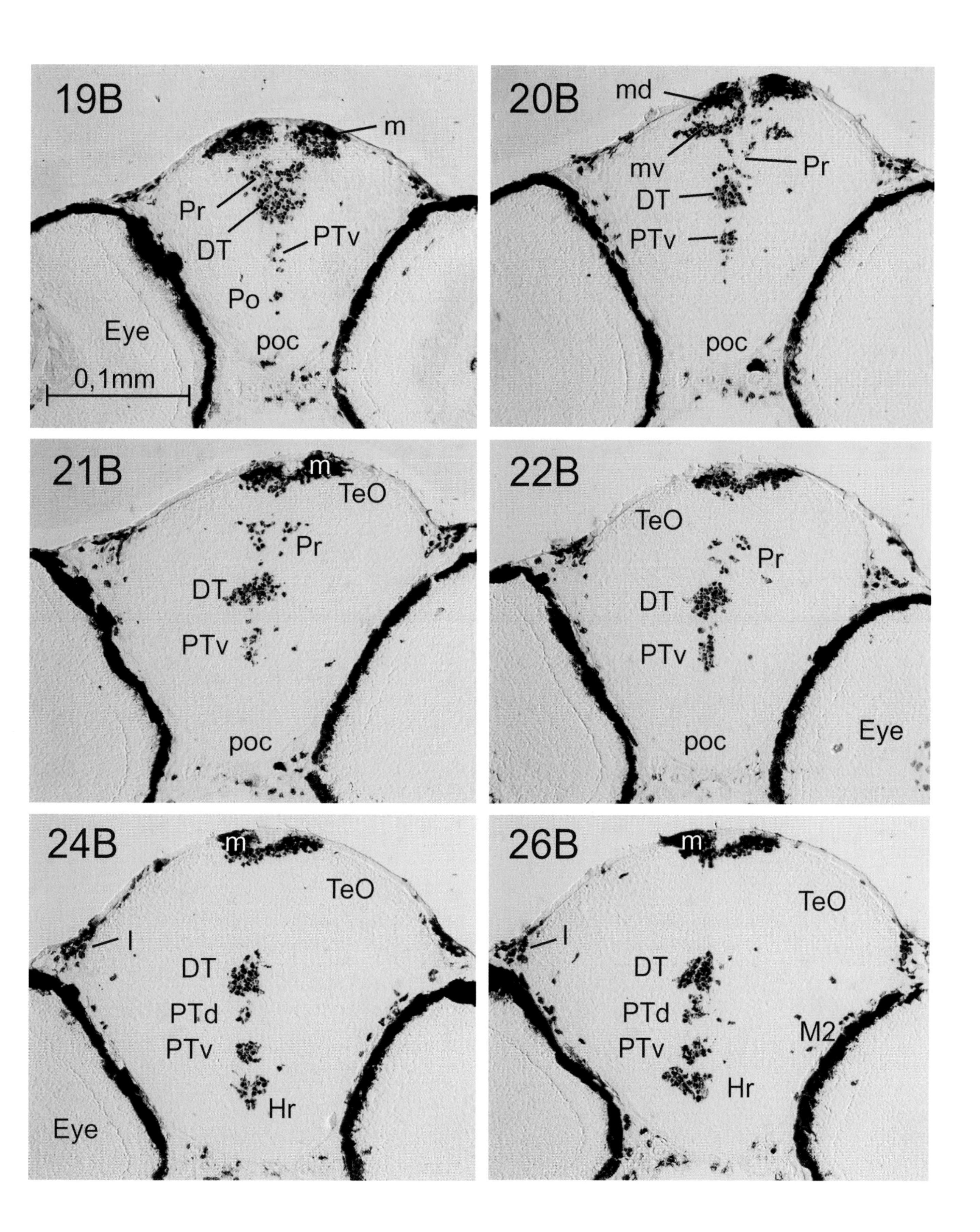
19B
m
Pr
DT
PTv
Po
Eye
poc
0,1mm
20B
md
mv
Pr
DT
PTv
poc
21B
m
TeO
Pr
DT
PTv
poc
22B
TeO
Pr
DT
PTv
poc
Eye
24B
m
TeO
l
DT
PTd
PTv
Hr
Eye
26B
m
TeO
l
DT
PTd
PTv
M2
Hr

BrdU (continued, 4th plate)

BrdU Cells in the 5-Day Zebrafish Brain

Description of levels on facing page

CNS: Large ventricular BrdU populations are present in intermediate hypothalamus (characterized by lateral ventricular recess) and caudal hypothalamus (characterized by posterior ventricular recess; see horizontal view below, right side). The hypothalamus forms the ventral part of the inferior lobe (interpreted as basal plate of anterior forebrain; i.e., secondary prosencephalon). The proliferation of the ventral posterior tuberculum extends into the dorsal portion of the inferior lobe (interpreted as basal plate of ventral thalamic prosomere/P3). Very few BrdU-positive cells are seen in the synencephalic region of the nucleus of the medial longitudinal fascicle (basal plate of P1), but towards the midbrain tegmentum and (medullary) midbrain–hindbrain boundary, BrdU-positive cells are seen ventral to the valvula cerebelli, which itself contains large BrdU-positive zones. Medial and lateral tectal BrdU-positive zones continue at these levels; a new one emerges in torus semicircularis in most caudolateral midbrain roof.

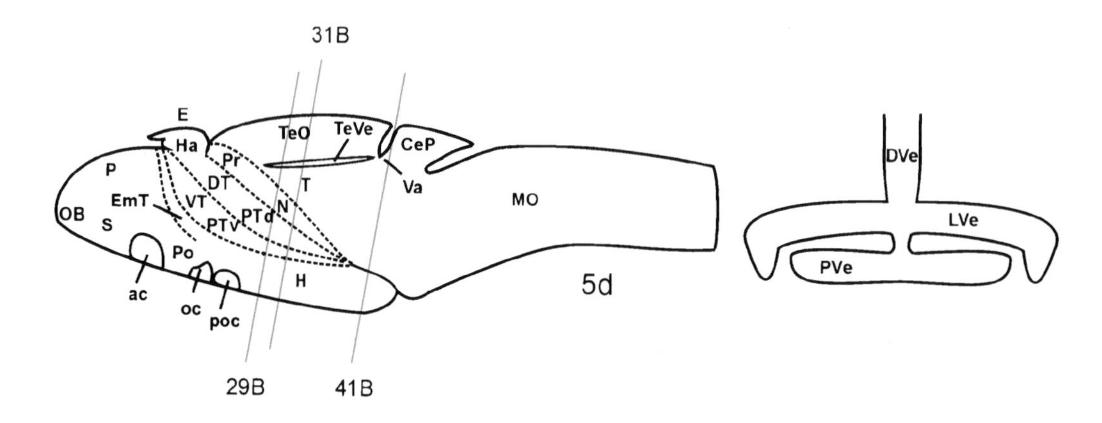

Abbreviations

ac	anterior commissure	OB	olfactory bulb
CeP	cerebellar plate	oc	optic chiasma
DT	dorsal thalamus (thalamus)	P	pallium
DVe	diencephalic ventricle	Po	preoptic region
E	epiphysis	poc	postoptic commissure
EmT	eminentia thalami	Pr	pretectum
H	hypothalamus	PTd	dorsal part of posterior tuberculum
Ha	habenula	PTv	ventral part of posterior tuberculum
Hc	caudal hypothalamus	PVe	posterior ventricular recess of H
Hi	intermediate hypothalamus	S	subpallium
l	lateral tectal proliferation zone	T	midbrain tegmentum
LVe	lateral ventricular recess of H	TeO	tectum opticum
M2	migrated posterior tubercular area	TeVe	tectal ventricle
m	medial tectal proliferation zone	TS	torus semicircularis
MO	medulla oblongata	Va	valvula cerebelli
N	region of the nucleus of medial longitudinal fascicle	VT	ventral thalamus (prethalamus)

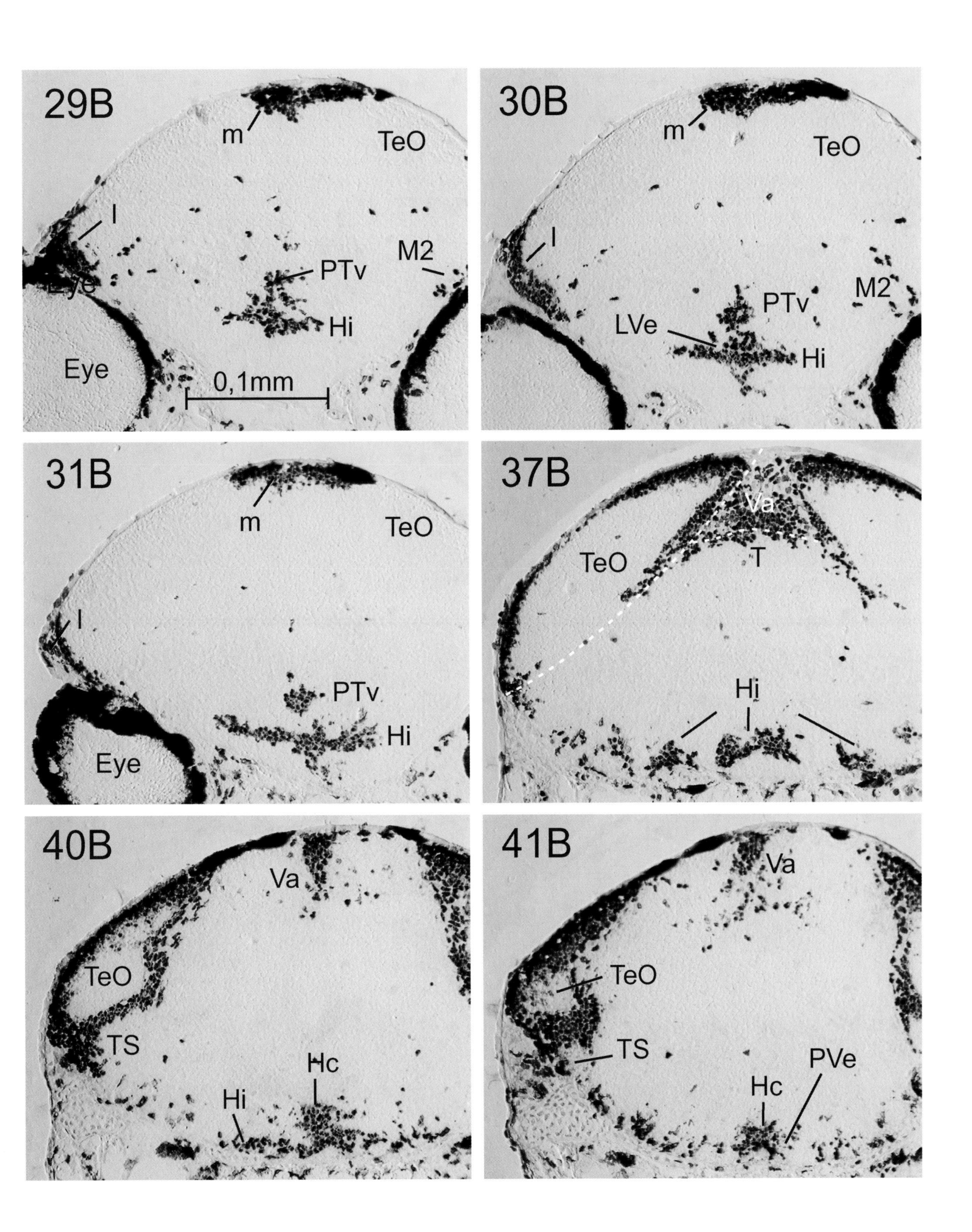
29B
m
TeO
I
M2
PTv
Hi
Eye
Eye
0,1mm
30B
m
TeO
I
PTv
M2
LVe
Hi
31B
m
TeO
I
PTv
Hi
Eye
37B
Va
TeO
T
Hi
40B
Va
TeO
TS
Hc
Hi
41B
Va
TeO
TS
Hc
PVe

BrdU (continued, 5th plate)

BrdU Cells in 5-Day Zebrafish Brain

Description of levels on facing page

CNS: The cerebellar plate displays BrdU-positive cells in the medial and ventral proliferative zones, as well as in the external granular layer. There are also distinct BrdU populations in the eminentia granularis. The rhombic lip (rim of dorsal opening of medulla oblongata, the rhombic fossa) contains many BrdU-positive cells. In contrast, the remainder of the medulla oblongata shows few BrdU cells at all levels.

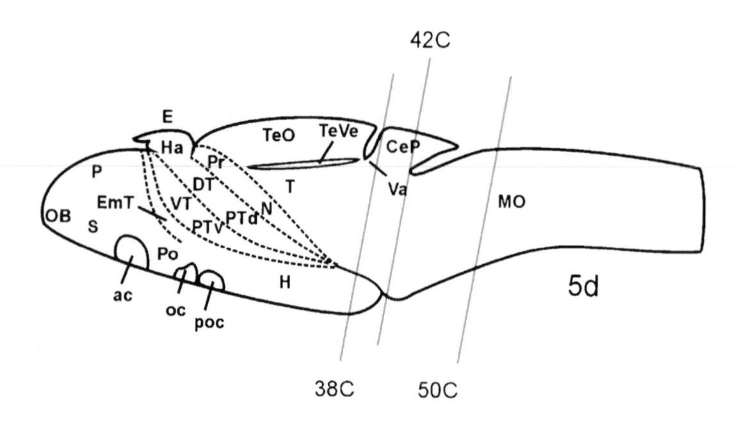

Abbreviations

ac	anterior commissure	oc	optic chiasma
CeP	cerebellar plate	P	pallium
Ch	chorda dorsalis	Po	preoptic region
DT	dorsal thalamus (thalamus)	poc	postoptic commissure
E	epiphysis	Pr	pretectum
EG	eminentia granularis	PTd	dorsal part of posterior tuberculum
EGL	external granular layer		
EmT	eminentia thalami	PTv	ventral part of posterior tuberculum
H	hypothalamus		
Ha	habenula	RL	rhombic lip
Hc	caudal hypothalamus	S	subpallium
LHP	lateral hinge-point between cerebellum and medulla oblongata	T	midbrain tegmentum
		TeO	tectum opticum
MCP	medial cerebellar proliferation	TeVe	tectal ventricle
MO	medulla oblongata	TS	torus semicircularis
N	region of the nucleus of medial longitudinal fascicle	Va	valvula cerebelli
		VCP	ventral cerebellar proliferation
OB	olfactory bulb	VT	ventral thalamus (prethalamus)

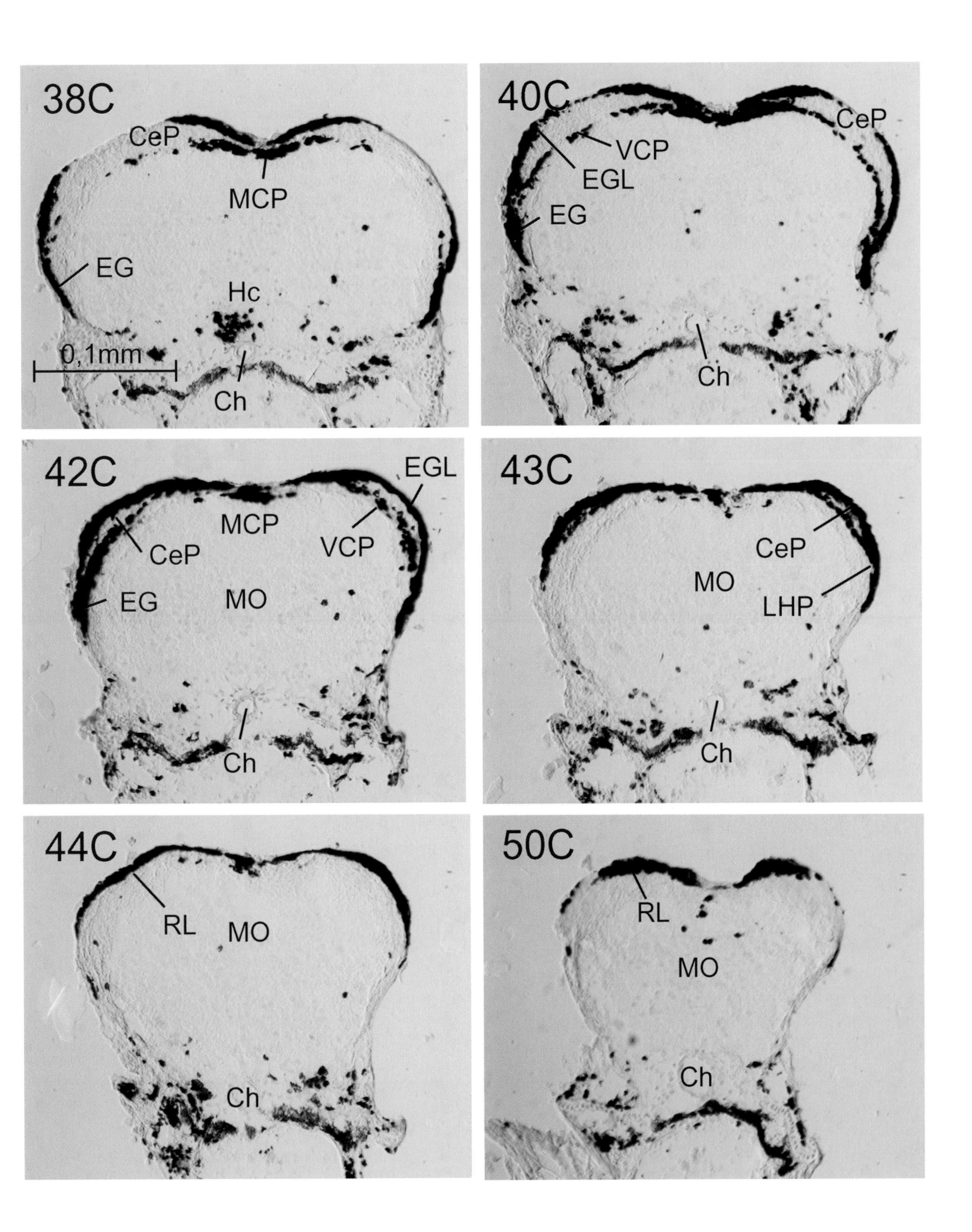

38C
CeP
MCP
EG
Hc
0,1mm
Ch
40C
CeP
VCP
EGL
EG
Ch
42C
EGL
MCP
VCP
CeP
EG
MO
Ch
43C
CeP
MO
LHP
Ch
44C
RL
MO
Ch
50C
RL
MO
Ch

Hu-proteins (shown with anti-Hu monoclonal antibody 16A11; source: Monoclonal Antibody Facility, Eugene, OR).
Description: mRNA binding proteins with a gene-regulatory function at the posttranscriptional level (Marusich et al. 1994). Antibody used recognizes peptide sequence present in various Hu-proteins, including those specifically expressed in neuronal cells (i.e., HuD, HuC; Park et al. 2000). Expression in early differentiating neuronal cells.

Protein Expression Domains in the 5-Day Zebrafish Brain

General appearance: The distribution of Hu-positive cells has not changed much compared to 3 days: although most CNS and PNS cells are Hu positive, regions with ongoing strong proliferation still show less, more weakly stained Hu-positive cells compared with more mature regions, and the ventricularly located proliferation zones remain Hu free (compare with BrdU at 5 days). As at 3 days, some clearly postmitotic, differentiated areas remain Hu negative (e.g., retinal outer nuclear layer, subcommissural organ, lateral torus, diffuse nucleus of inferior lobe, some cells in M2, Mauthner cells, rhombencephalic floor plate cells).

Description of levels on facing page

Sensory organs: Many strongly Hu-positive cells are seen in the olfactory epithelium. Many Hu-positive cells are seen in ganglionic and inner nuclear layers of retina, but not in outer nuclear layer and retinal peripheral edge.

CNS: In the telencephalon, many Hu-stained cells are present in the olfactory bulb (which exhibits now many more glomeruli, i.e., globular neuropil structures formed both by axonal arborizations of olfactory epithelial sensory cells and by dendrites of olfactory bulb neurons), as well as in the gray matter of pallium and (ventral and dorsal divisions) of subpallium. Subpallial and pallial Hu domains lie at least one cell row away from the ventricular surface. Note strongly stained Hu cells in M4, an early migrated cell aggregate possibly representing the future lateral nucleus of the ventral telencephalic area.

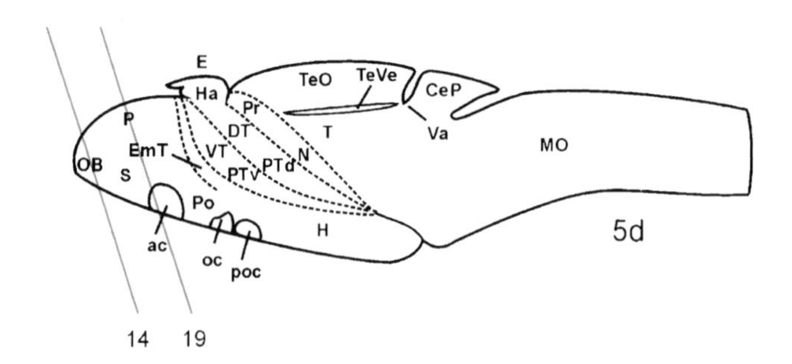

Abbreviations

ac	anterior commissure	P	pallium
CeP	cerebellar plate	Po	preoptic region
DT	dorsal thalamus (thalamus)	poc	postoptic commissure
E	epiphysis	Pr	pretectum
EmT	eminentia thalami	PTd	dorsal part of posterior tuberculum
Gl	olfactory bulb glomeruli	PTv	ventral part of posterior tuberculum
H	hypothalamus	S	subpallium
Ha	habenula	Sd	dorsal division of S
M4	telencephalic migrated area	Sv	ventral division of S
MO	medulla oblongata	T	midbrain tegmentum
N	region of the nucleus of medial longitudinal fascicle	TeO	tectum opticum
		TVe	telencephalic ventricle
OB	olfactory bulb	TeVe	tectal ventricle
oc	optic chiasma	Va	valvula cerebelli
OE	olfactory epithelium	VT	ventral thalamus (prethalamus)

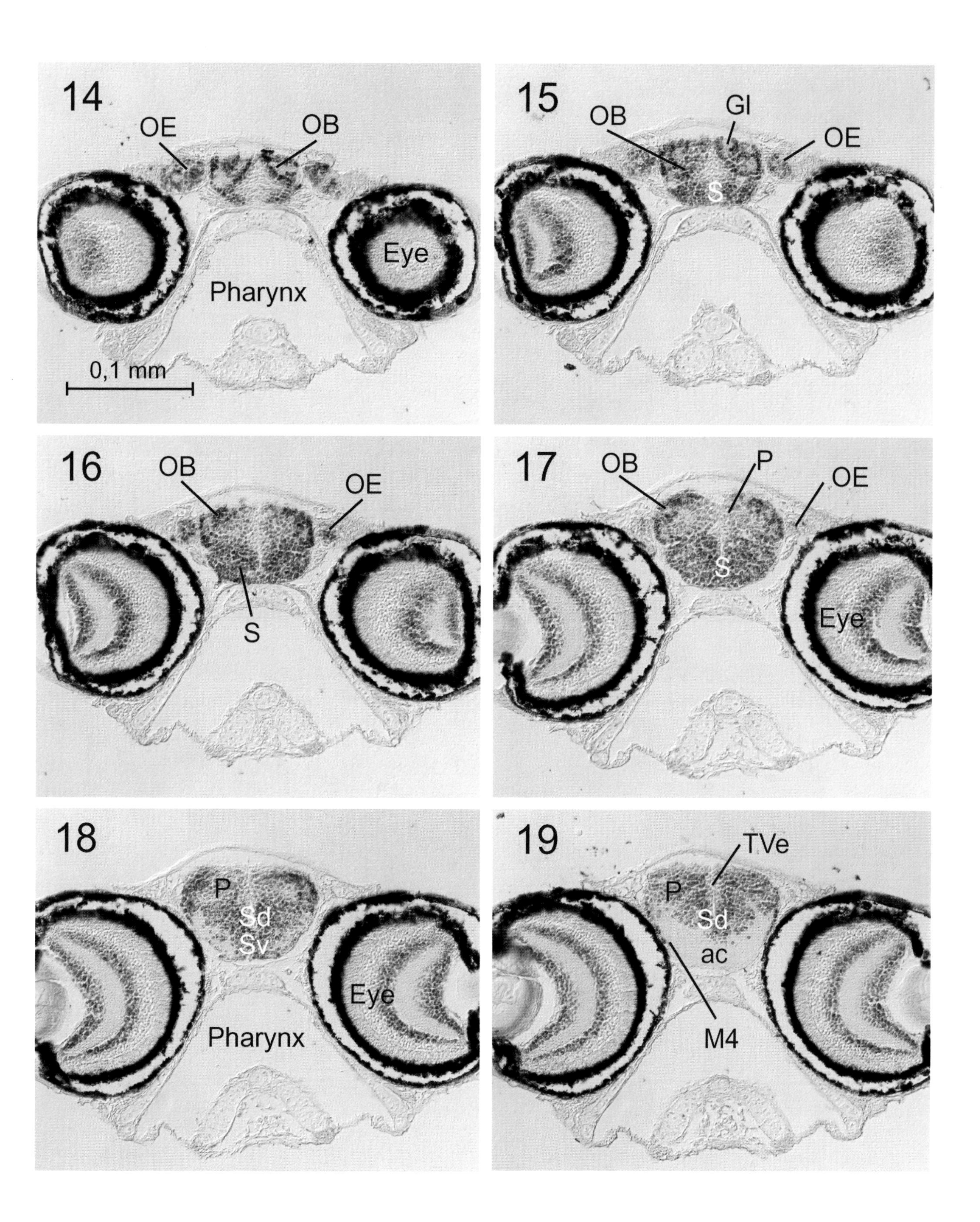
14
OE
OB
Eye
Pharynx
0,1 mm
15
OB
Gl
OE
S
16
OB
OE
S
17
OB
P
OE
S
Eye
18
P
Sd
Sv
Eye
Pharynx
19
TVe
P
Sd
ac
M4

Hu-proteins (continued, 2nd plate)

Protein Expression Domains in the 5-Day Zebrafish Brain

Description of levels on facing page

Sensory organs: Many Hu-positive cells are seen in ganglionic and inner nuclear layers of retina, but not in outer nuclear layer and retinal peripheral edge.

CNS: Telencephalic Hu-stained cells are present in gray matter of pallium and subpallium (dorsal division visible on facing page continues dorsally to anterior commissure, i.e., at supracommissural levels). Note particularly strongly stained Hu cells in M4, an early migrated cell aggregate possibly representing the future lateral nucleus of the ventral telencephalic area. Telencephalic Hu domains lie at least one cell row away from the ventricular surface. Note that some migrated (non-proliferative) pallial cells are Hu negative. Posterior to the anterior commissure, the preoptic region emerges and shows a greater proportion of Hu-positive cells compared with 3 days, but remains strongly proliferative and Hu negative in its central part. The eminentia thalami and its most anterolaterally migrated part (i.e., the future entopeduncular complex, M3) contains strongly Hu-positive cells.

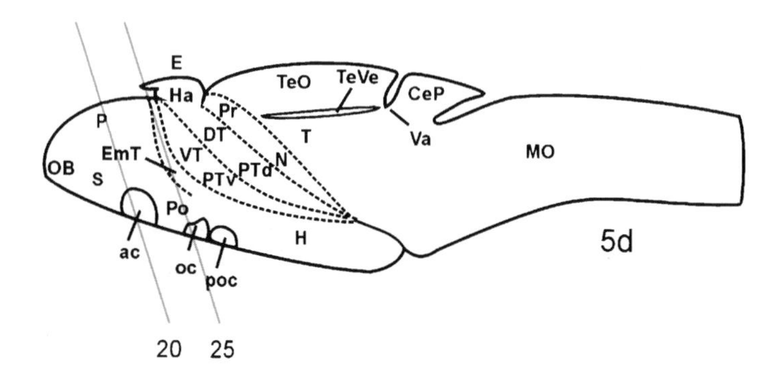

Abbreviations

ac	anterior commissure	P	pallium
CeP	cerebellar plate	Po	preoptic region
DT	dorsal thalamus (thalamus)	poc	postoptic commissure
E	epiphysis	Pr	pretectum
EmT	eminentia thalami	PTd	dorsal part of posterior tuberculum
H	hypothalamus	PTv	ventral part of posterior tuberculum
Ha	habenula		
lfb	lateral forebrain bundle	S	subpallium
M3	migrated area of EmT	Sd	dorsal division of S
M4	telencephalic migrated area	Sv	ventral division of S
MO	medulla oblongata	T	midbrain tegmentum
N	region of the nucleus of medial longitudinal fascicle	TeO	tectum opticum
		TVe	telencephalic ventricle
OB	olfactory bulb	TeVe	tectal ventricle
oc	optic chiasma	Va	valvula cerebelli
on	optic nerve	VT	ventral thalamus (prethalamus)

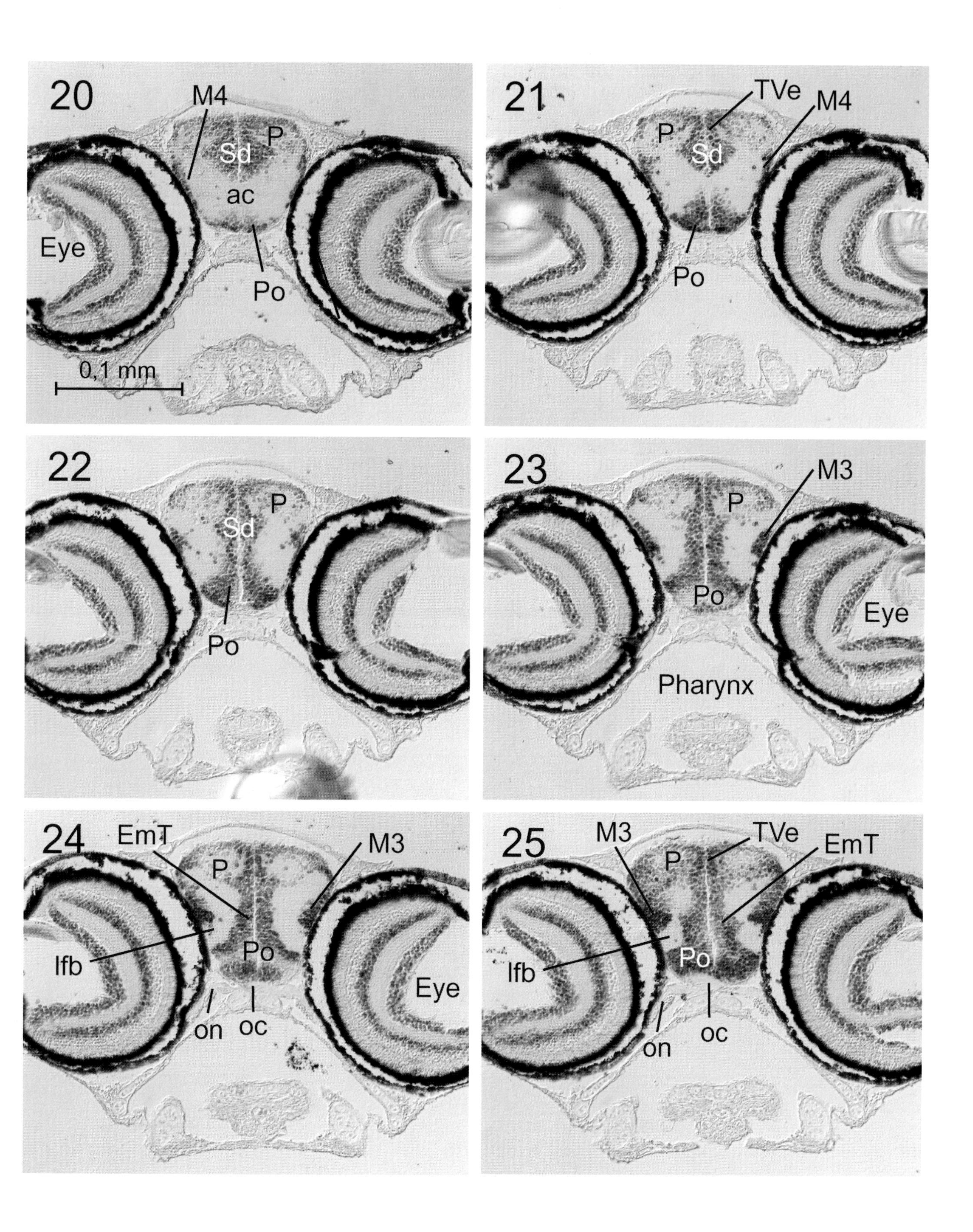
20
M4
P
Sd
ac
Eye
Po
0,1 mm
21
TVe
M4
P
Sd
Po
22
P
Sd
Po
23
M3
P
Po
Eye
Pharynx
24
EmT
M3
P
lfb
Po
Eye
on
oc
25
M3
TVe
EmT
P
lfb
Po
on
oc

Hu-proteins (continued, 3rd plate)

Protein Expression Domains in the 5-Day Zebrafish Brain

Description of levels on facing page

Sensory organs: Many Hu-positive cells are seen in ganglionic and inner nuclear layers of retina, but not in outer nuclear layer and retinal peripheral edge.

CNS: The preoptic region shows many Hu cells at its caudal pole. At these levels, a distinct Hu-positive cell stream visibly arches around the lateral forebrain bundle in the area of the eminentia thalami (future entopeduncular complex, M3). In the diencephalon, Hu-stained cells are present in the habenula and epiphysis, but not in the (postmitotic) subcommissural organ. The (strongly proliferative) dorsal and ventral thalami remain Hu free at rostral levels, but increasingly more Hu-positive cells emerge at more caudal thalamic levels. Also, Hu cells are abundant in the peripherally migrated pretectal region (future superficial pretectum, M1) and the migrated posterior tubercular region M2 (future preglomerular complex; now separated from the more medial gray matter). Caudal to the postoptic commissure, the emerging rostral hypothalamus exhibits many Hu-positive cells.

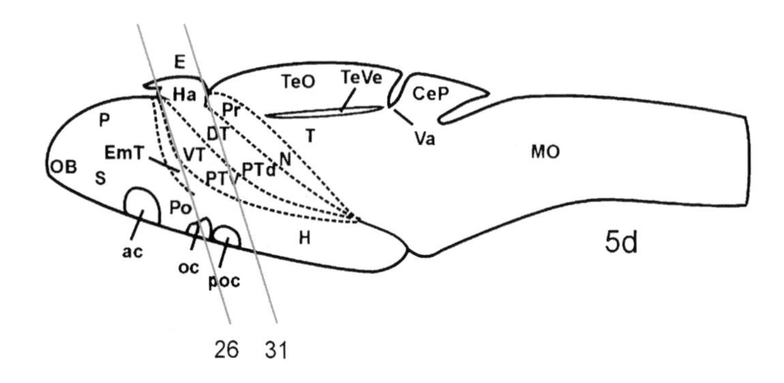

Abbreviations

ac anterior commissure
CeP cerebellar plate
DT dorsal thalamus (thalamus)
E epiphysis
EmT eminentia thalami
H hypothalamus
Ha habenula
Hr rostral hypothalamus
lfb lateral forebrain bundle
M1 migrated pretectal area
M2 migrated posterior tubercular area
M3 migrated area of EmT
MO medulla oblongata
N region of the nucleus of medial longitudinal fascicle
OB olfactory bulb
oc optic chiasma
P pallium
Pi pigment
Po preoptic region
poc postoptic commissure
Pr pretectum
PT posterior tuberculum
PTd dorsal part of posterior tuberculum
PTv ventral part of posterior tuberculum
S subpallium
SCO subcommissural organ
T midbrain tegmentum
TeO tectum opticum
TeVe tectal ventricle
Va valvula cerebelli
VT ventral thalamus (prethalamus)

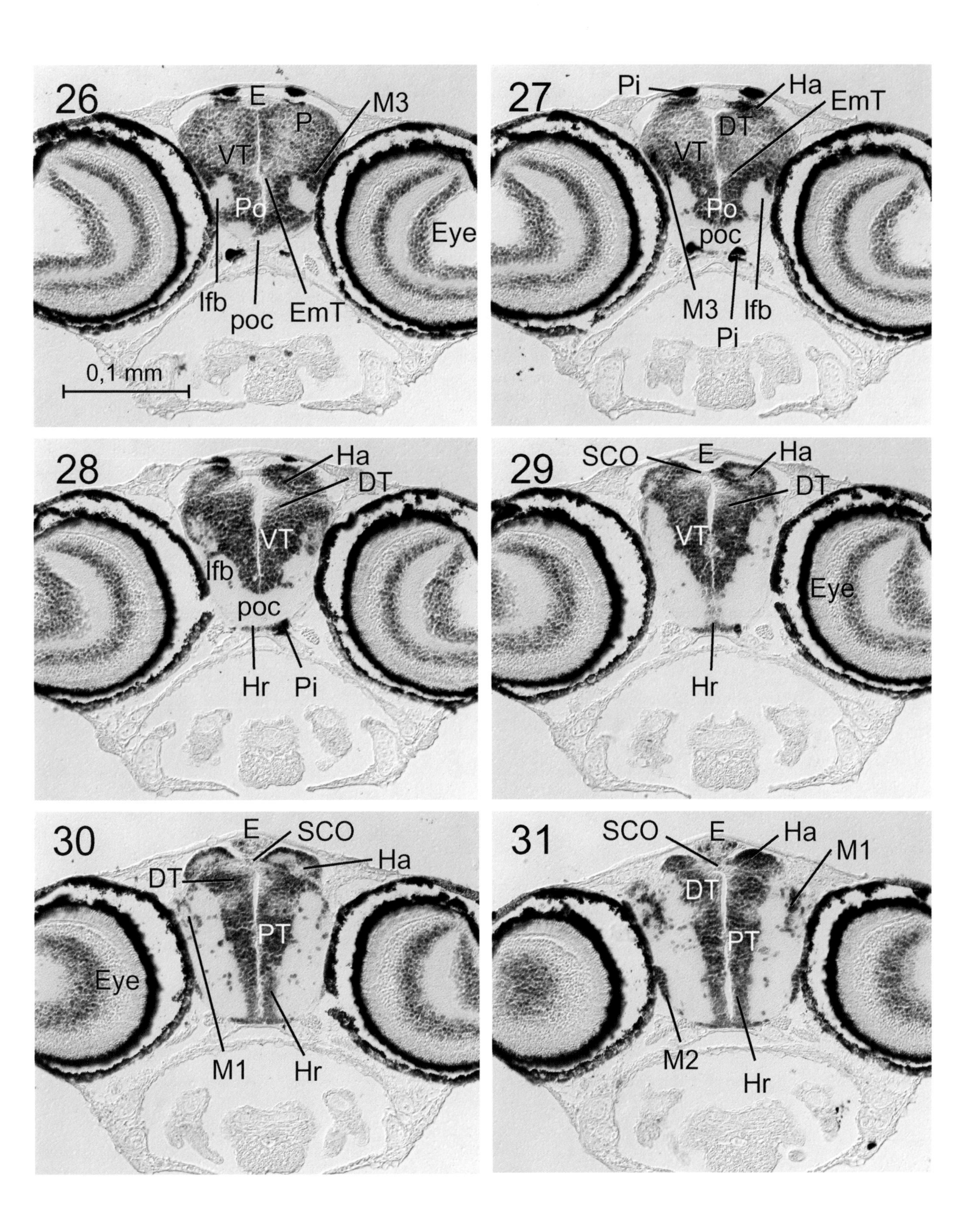
26
E
P
M3
VT
Po
Eye
lfb
poc
EmT
0,1 mm
27
Pi
Ha
EmT
DT
VT
Po
poc
M3
lfb
Pi
28
Ha
DT
VT
lfb
poc
Hr
Pi
29
SCO
E
Ha
DT
VT
Eye
Hr
30
E
SCO
Ha
DT
PT
Eye
M1
Hr
31
SCO
E
Ha
M1
DT
PT
M2
Hr

Hu-proteins (continued, 4th plate)

Protein Expression Domains in the 5-Day Zebrafish Brain

Description of levels on facing page

Sensory organs: Many Hu-positive cells are seen in ganglionic and inner nuclear layers of retina, but not in outer nuclear layer and retinal peripheral edge.

CNS: In the diencephalon, many Hu-stained cells are present in pretectum (including its migrated region M1; future superficial pretectum) and dorsal thalamus proper, as well as in the posterior tuberculum (characterized by its less laterally extending gray matter compared to dorsal thalamus; dorsal and ventral posterior tubercular divisions hard to distinguish). Note the peripheral Hu-positive cells in the migrated posterior tubercular region M2 (future preglomerular complex; separated from the more medial gray matter). Rostral and intermediate hypothalami exhibit many Hu-positive cells. In the optic tectum, many more peripheral Hu-positive cells are seen compared with 3 days (but note weaker staining than in adjacent tegmentum). Note strongly Hu-positive midline cells in most anterior tectum possibly representing the torus longitudinalis. The strongly stained Hu cells at the base of the anterior optic tectum possibly represent the griseum tectale, a part of the alar plate mesencephalon curving rostrally below the tectal ventricle; this region has many peripherally migrated cells dorsally adjacent to M1.

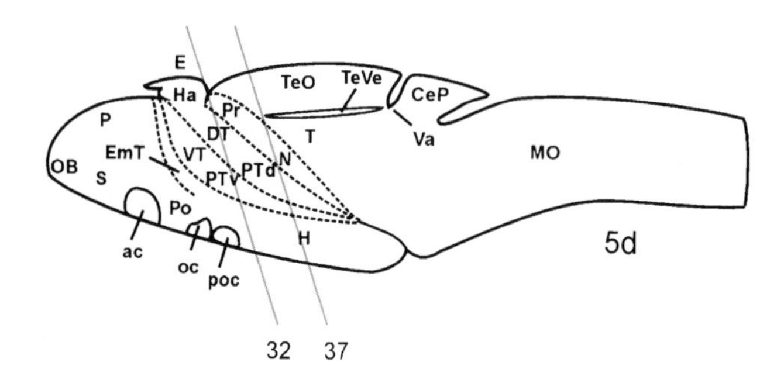

Abbreviations

ac	anterior commissure	OB	olfactory bulb
CeP	cerebellar plate	oc	optic chiasma
DIL	diffuse nucleus of inferior lobe	on	optic nerve
DT	dorsal thalamus (thalamus)	P	pallium
E	epiphysis	pc	posterior commissure
EmT	eminentia thalami	Pi	pigment
fr	fasciculus retroflexus	Po	preoptic region
GT	griseum tectale	poc	postoptic commissure
H	hypothalamus	Pr	pretectum
Ha	habenula	PT	posterior tuberculum
Hi	intermediate hypothalamus	PTd	dorsal part of posterior tuberculum
Hr	rostral hypothalamus	PTv	ventral part of posterior tuberculum
Hy	hypophysis (pituitary)	S	subpallium
lfb	lateral forebrain bundle	SCO	subcommissural organ
LVe	lateral ventricular recess of H	T	midbrain tegmentum
M1	migrated pretectal area	TeO	tectum opticum
M2	migrated posterior tubercular area	TeVe	tectal ventricle
MO	medulla oblongata	TL	torus longitudinalis
N	region of the nucleus of medial longitudinal fascicle	Va	valvula cerebelli
		VT	ventral thalamus (prethalamus)

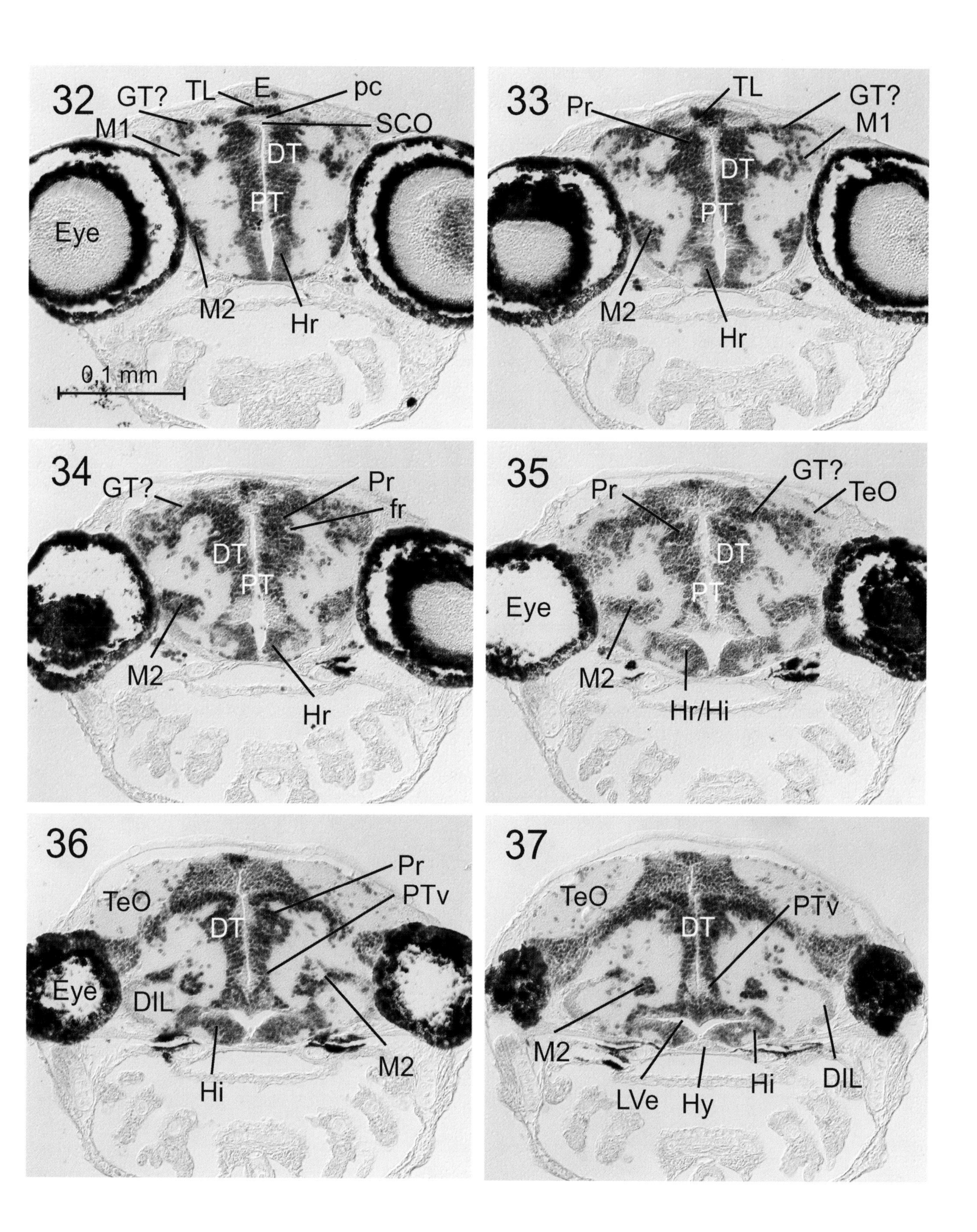

32
GT?
TL
E
pc
SCO
M1
DT
PT
Eye
M2
Hr
0,1 mm
33
Pr
TL
GT?
M1
DT
PT
M2
Hr
34
GT?
Pr
fr
DT
PT
M2
Hr
35
Pr
GT?
TeO
DT
PT
Eye
M2
Hr/Hi
36
Pr
PTv
TeO
DT
Eye
DIL
M2
Hi
37
TeO
PTv
DT
M2
LVe
Hy
Hi
DIL

Hu-proteins (continued, 5th plate)

Protein Expression Domains in the 5-Day Zebrafish Brain

Description of levels on facing page

Sensory organs/PNS: Many darkly stained Hu cells are present in various cranial nerve ganglia.

CNS: At these levels, the diencephalon exhibits Hu-stained cells in the area of the nucleus of the medial longitudinal fascicle (basal synencephalon), in the posterior tuberculum (primarily ventral division visible on facing page, extending into dorsal part of inferior lobe). Note peripheral Hu-positive (but also many Hu-negative, but postmitotic) cells in the migrated posterior tubercular region M2 (future preglomerular complex). The intermediate hypothalamus (characterized by the lateral recess ventricle) has many Hu-positive cells (more anteriorly, less posteriorly). At these inferior lobe levels, the caudal hypothalamus (Hu positive) replaces the ventral part of the posterior tuberculum. The optic tectum has many more Hu-positive cells; the torus semicircularis is slightly stronger Hu positive. Caudal to the basal synencephalon, the basal plate midbrain tegmentum appears and contains many strongly stained Hu cells as does the caudally adjacent medulla oblongata. Note well differentiated oculomotor nerve nucleus and interpeduncular nucleus (both Hu positive) and Hu-negative floor plate cells at the bottom of rhombencephalic ventricle. The hypophysis is free of a Hu signal. Torus lateralis and diffuse nucleus of inferior lobe are clearly newly recognizable and are Hu negative.

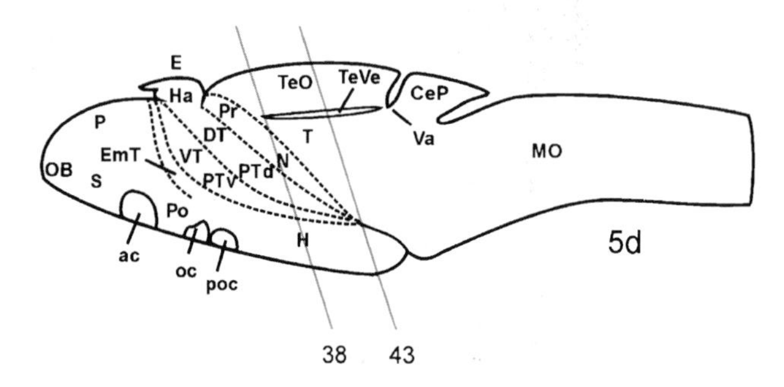

Abbreviations

ac	anterior commissure	NIn	nucleus interpeduncularis
ALLG	anterior lateral line ganglion	OB	olfactory bulb
CeP	cerebellar plate	P	pallium
Ch	chorda dorsalis	oc	optic chiasma
DIL	diffuse nucleus of inferior lobe	Po	preoptic region
DT	dorsal thalamus (thalamus)	poc	postoptic commissure
E	epiphysis	Pr	pretectum
EmT	eminentia thalami	PTd	dorsal part of posterior tuberculum
FP	floor plate		
H	hypothalamus	PTv	ventral part of posterior tuberculum
Hc	caudal hypothalamus		
Hi	intermediate hypothalamus	PVe	posterior ventricular recess of H
Hr	rostral hypothalamus	S	subpallium
Hy	hypophysis (pituitary)	T	midbrain tegmentum
LVe	lateral ventricular recess of H	TeO	tectum opticum
M2	migrated posterior tubercular area	TeVe	tectal ventricle
mlf	medial longitudinal fascicle	TG	trigeminal ganglion
MO	medulla oblongata	TLa	torus lateralis
NIII	oculomotor nerve nucleus	TS	torus semicircularis
NIIIn	oculomotor nerve	Va	valvula cerebelli
N	region of the nucleus of medial longitudinal fascicle	VT	ventral thalamus (prethalamus)

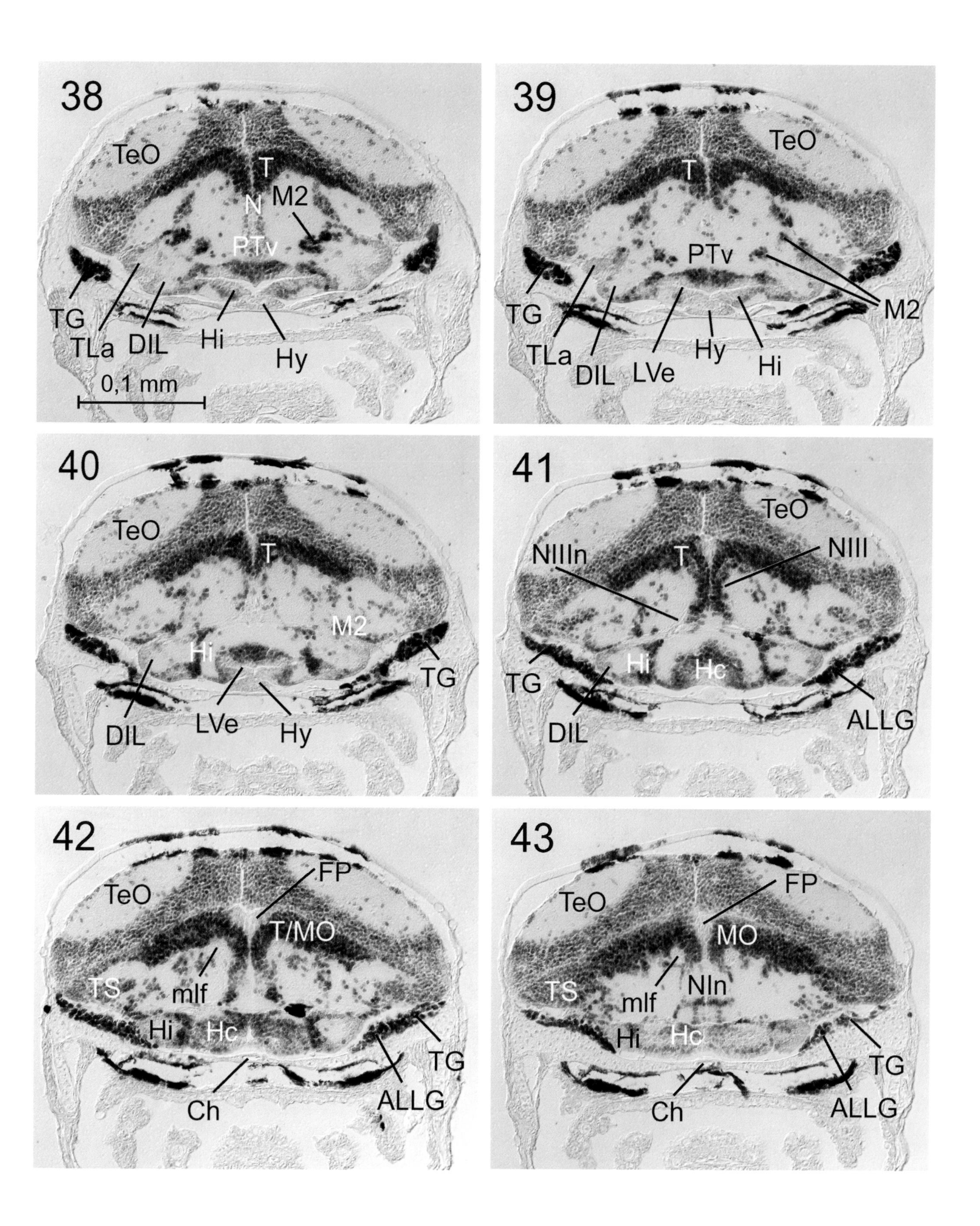
38
TeO
T
N
M2
PTv
TG
TLa
DIL
Hi
Hy
0,1 mm
39
TeO
T
PTv
M2
TG
TLa
DIL
LVe
Hy
Hi
40
TeO
T
M2
Hi
TG
DIL
LVe
Hy
41
TeO
NIIIn
T
NIII
Hi
Hc
TG
DIL
ALLG
42
TeO
FP
T/MO
mlf
TS
Hi
Hc
TG
Ch
ALLG
43
TeO
FP
MO
mlf
NIn
TS
Hi
Hc
TG
Ch
ALLG

Hu-proteins (continued, 6th plate)

Protein Expression Domains in the 5-Day Zebrafish Brain

Description of levels on facing page

Sensory organs/PNS: Many darkly stained Hu cells are present in various cranial nerve ganglia.

CNS: At these levels, the caudal hypothalamus (characterized by the posterior recess ventricle) continues into the most caudal part of the inferior lobe where it also replaces the intermediate hypothalamus. The caudal hypothalamus exhibits many Hu-positive cells (especially anteriorly, much less posteriorly). Many weakly Hu-positive cells are present in the optic tectum and torus semicircularis. Also the valvula cerebelli contains more, strongly positive Hu cells compared with 3 days. The cerebellar plate and especially the eminentia granularis contain now many, more strongly stained Hu-positive cells than at 3 days. The medulla oblongata contains many strongly stained Hu cells at otic levels, as do some identifiably differentiated medullary structures, namely interpeduncular nucleus, superior raphe and superior reticular formation

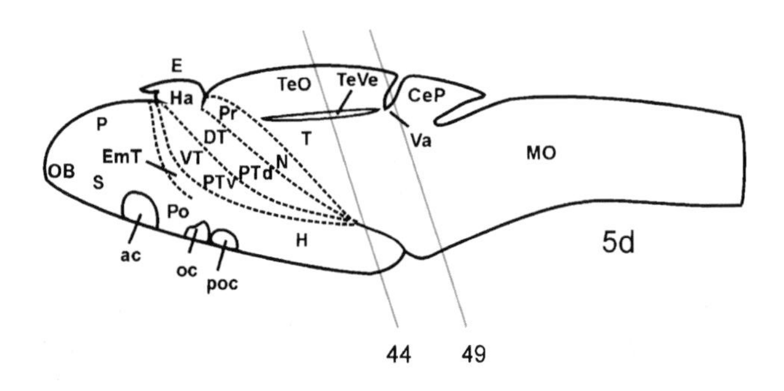

Abbreviations

ac	anterior commissure
ALLG	anterior lateral line ganglion
cec	cerebellar commissure
CeP	cerebellar plate
Ch	chorda dorsalis
DT	dorsal thalamus (thalamus)
E	epiphysis
EG	eminentia granularis
EmT	eminentia thalami
FG	facial ganglion
H	hypothalamus
Hc	caudal hypothalamus
LVe	lateral ventricular recess of H
mlf	medial longitudinal fascicle
MO	medulla oblongata
NIII	oculomotor nerve nucleus
N	region of the nucleus of medial longitudinal fascicle
NIn	nucleus interpeduncularis
OB	olfactory bulb
oc	optic chiasma
OC	otic capsule
OG	octaval ganglion
P	pallium
Pi	pigment
Po	preoptic region
poc	postoptic commissure
Pr	pretectum
PTd	dorsal part of posterior tuberculum
PTv	ventral part of posterior tuberculum
PVe	posterior ventricular recess of H
RVe	rhombencephalic ventricle
S	subpallium
SR	superior raphe
SRF	superior reticular formation
T	midbrain tegmentum
TeO	tectum opticum
TeVe	tectal ventricle
TS	torus semicircularis
Va	valvula cerebelli
VT	ventral thalamus (prethalamus)

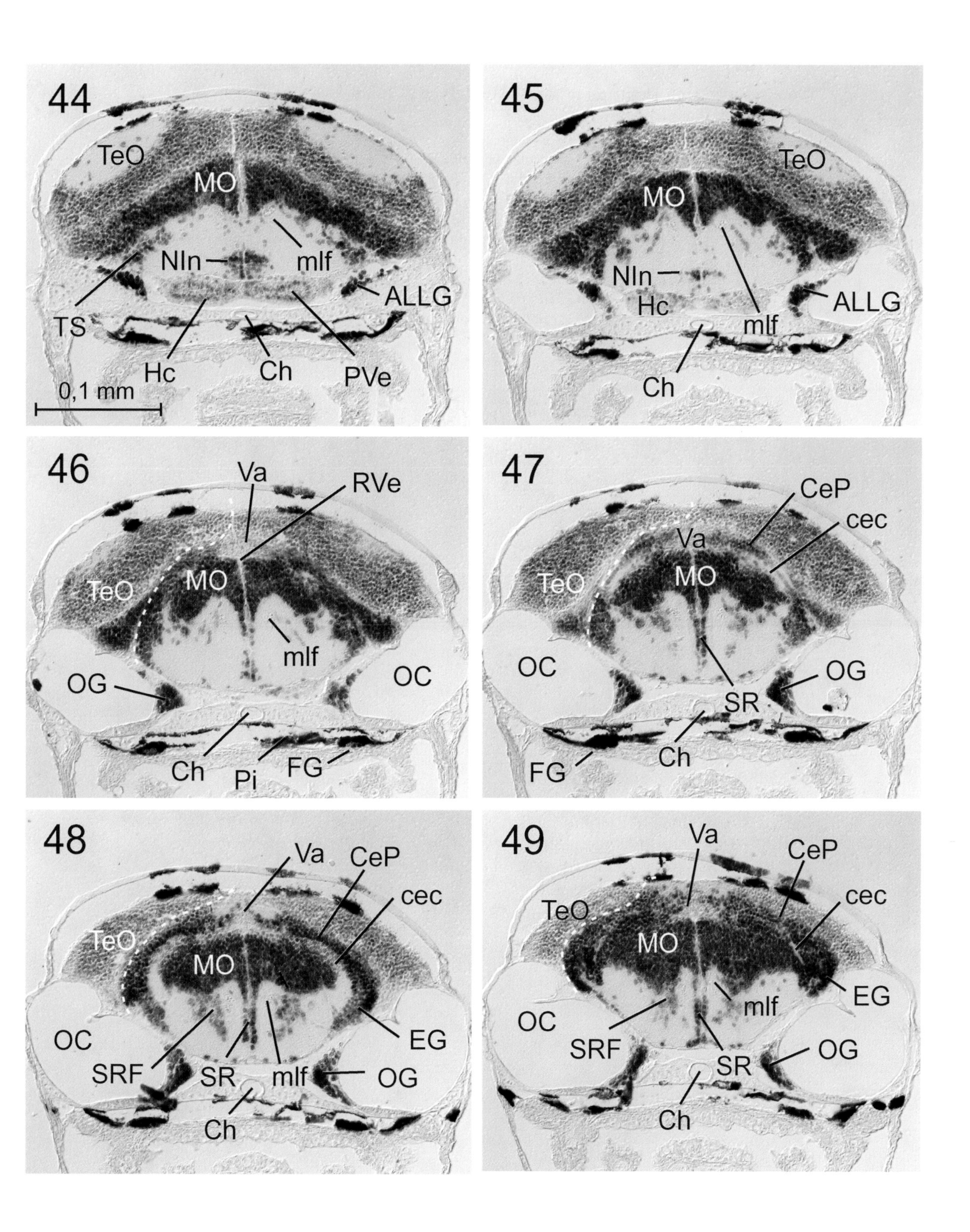
44
TeO
MO
NIn
mlf
ALLG
TS
Hc
Ch
PVe
0,1 mm
45
TeO
MO
NIn
Hc
mlf
ALLG
Ch
46
Va
RVe
TeO
MO
mlf
OG
OC
Ch
Pi
FG
47
CeP
cec
Va
TeO
MO
OC
OG
SR
Ch
FG
48
Va
CeP
cec
TeO
MO
OC
EG
SRF
SR
mlf
OG
Ch
49
Va
CeP
cec
TeO
MO
EG
mlf
OC
SRF
SR
OG
Ch

Hu-proteins (continued, 7th plate)

Protein Expression Domains in the 5-Day Zebrafish Brain

Description of levels on facing page

Sensory organs/PNS: Many darkly stained Hu cells are present in various cranial nerve ganglia.

CNS: As at 3 days, the cerebellar plate displays fewer Hu cells in the caudal part, clearly sandwiched between the most superficial area of the proliferative external granular layer and the basal proliferative zone (compare with BrdU at 5 days). The eminentia granularis contains many strongly stained Hu-positive cells, as does the medulla oblongata at these otic levels, as well as some identifiably differentiated medullary structures, namely the superior raphe and superior/intermediate reticular formation. The differentiated Mauthner neuron appears to remain Hu free.

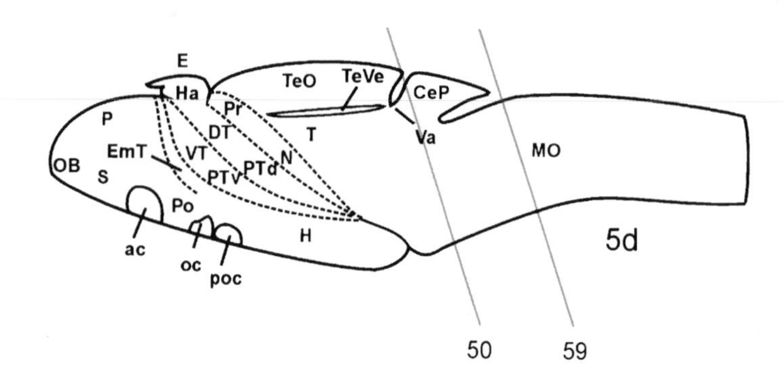

Abbreviations

ac	anterior commissure	OC	otic capsule
CC	cerebellar crest	OG	otic ganglion
CeP	cerebellar plate	P	pallium
Ch	chorda dorsalis	Pi	pigment
DT	dorsal thalamus (thalamus)	Po	preoptic region
E	epiphysis	poc	postoptic commissure
EG	eminentia granularis	Pr	pretectum
EmT	eminentia thalami	PTd	dorsal part of posterior tuberculum
GG	glossopharyngeal ganglion	PTv	ventral part of posterior tuberculum
H	hypothalamus	RVe	rhombencephalic ventricle
IMRF	intermediate reticular formation	S	subpallium
LHP	lateral hinge-point between cerebellum/medulla oblongata	SR	superior raphe
mlf	medial longitudinal fascicle	SRF	superior reticular formation
MN	Mauthner neuron	T	midbrain tegmentum
MO	medulla oblongata	TeO	tectum opticum
N	region of the nucleus of medial longitudinal fascicle	TeVe	tectal ventricle
OB	olfactory bulb	Va	valvula cerebelli
oc	optic chiasma	VT	ventral thalamus (prethalamus)

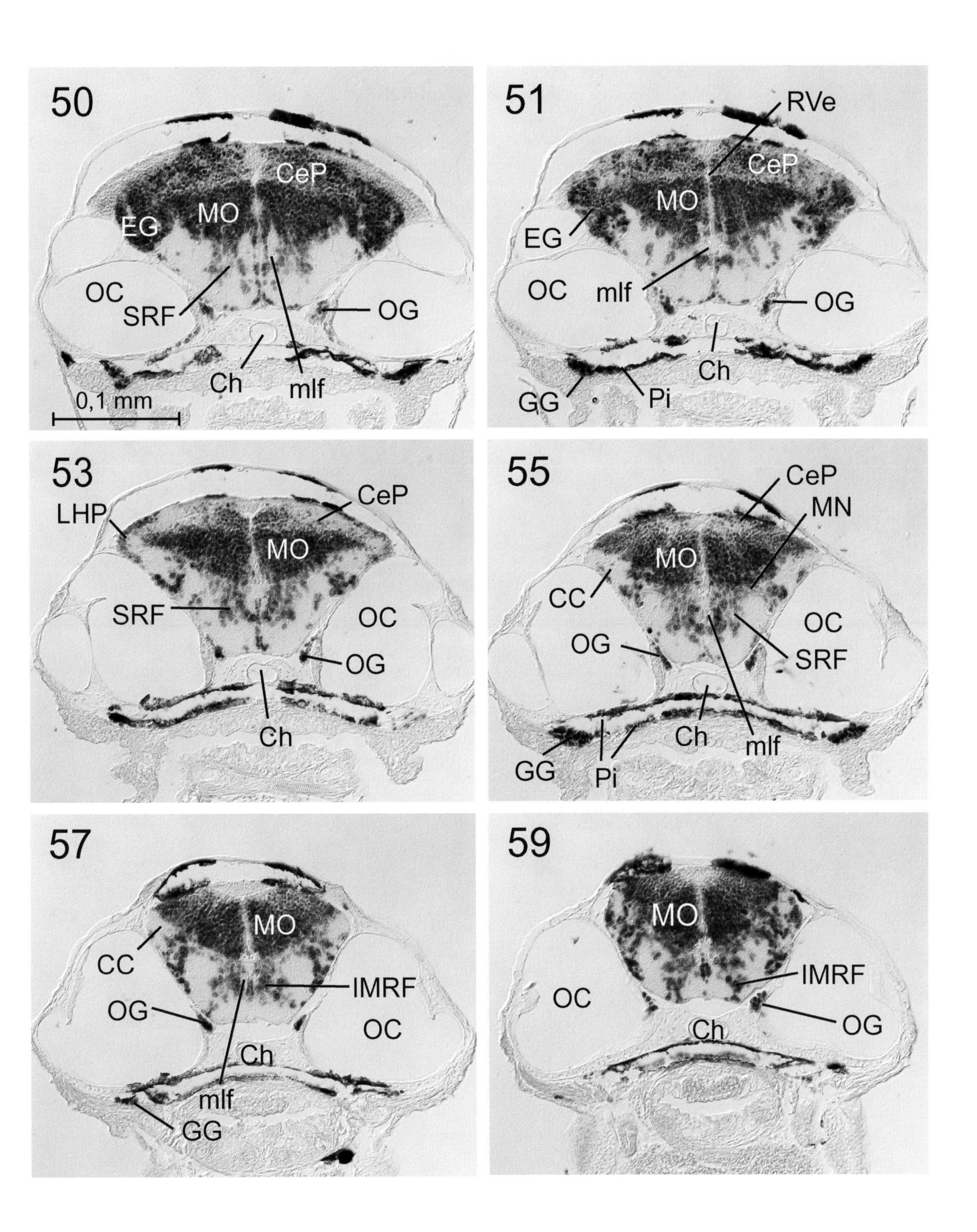
50
CeP
MO
EG
OC
SRF
OG
Ch
mlf
0,1 mm
51
RVe
CeP
MO
EG
OC
mlf
OG
Ch
GG
Pi
53
LHP
CeP
MO
SRF
OC
OG
Ch
55
CeP
MN
MO
CC
OC
OG
SRF
Ch
mlf
GG
Pi
57
MO
CC
IMRF
OG
OC
Ch
mlf
GG
59
MO
IMRF
OC
OG
Ch

Hu-proteins (continued, 8th plate)

Protein Expression Domains in the 5-Day Zebrafish Brain

Description of levels on facing page

Sensory organs/PNS: Many darkly stained Hu cells are present in various cranial nerve ganglia.

CNS: At postotic levels, the medulla oblongata continues to exhibit many strongly stained Hu cells, but the (proliferative) rhombic lip remains Hu free. In contrast to 3 days, many cells in the lateral medulla, appearing to derive from the rhombic lip, are now strongly Hu positive. Some identifiably differentiated medullary structures, namely lateral reticular nucleus, inferior olive, inferior raphe and inferior reticular formation are strongly Hu positive.

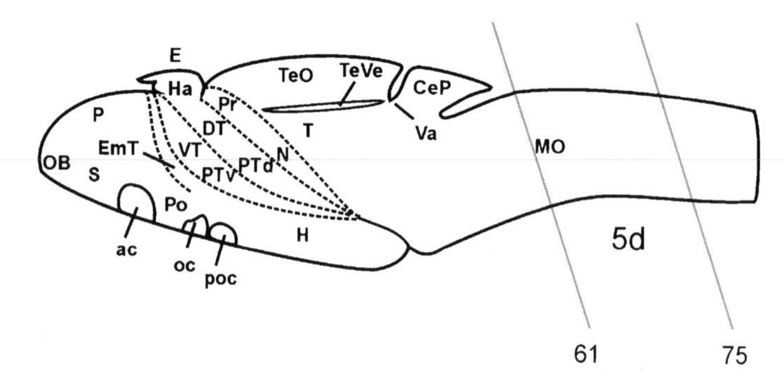

Abbreviations

ac	anterior commissure	P	pallium
CeP	cerebellar plate	Pi	pigment
Ch	chorda dorsalis	PLLG	posterior lateral line ganglion
DT	dorsal thalamus (thalamus)	Po	preoptic region
E	epiphysis	poc	postoptic commissure
EmT	eminentia thalami	Pr	pretectum
H	hypothalamus	PTd	dorsal part of posterior tuberculum
IO	inferior olive	PTv	ventral part of posterior tuberculum
IR	inferior raphe	RL	rhombic lip
IRF	inferior reticular formation	S	subpallium
LRN	lateral reticular nucleus	T	midbrain tegmentum
mlf	medial longitudinal fascicle	TeO	tectum opticum
MO	medulla oblongata	TeVe	tectal ventricle
N	region of the nucleus of medial longitudinal fascicle	Va	valvula cerebelli
OB	olfactory bulb	VG	vagal ganglion
oc	optic chiasma	VT	ventral thalamus (prethalamus)
OC	otic capsule		

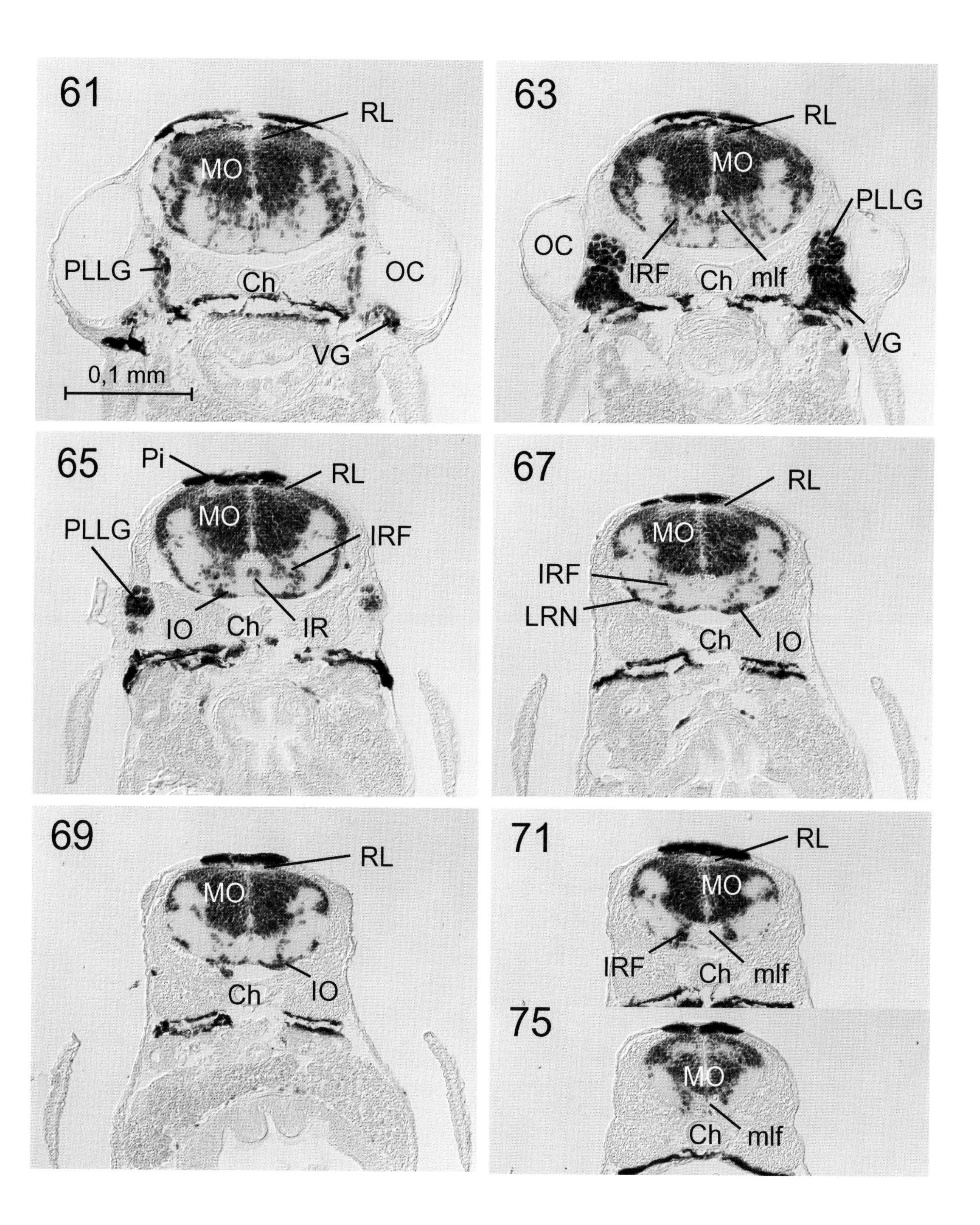
61
RL
MO
PLLG
Ch
OC
VG
0,1 mm
63
RL
MO
PLLG
OC
IRF
Ch
mlf
VG
65
Pi
RL
MO
PLLG
IRF
IO
Ch
IR
67
RL
MO
IRF
LRN
Ch
IO
69
RL
MO
Ch
IO
71
RL
MO
IRF
Ch
mlf
75
MO
Ch
mlf

Interpretation of Data—How to Use the Atlas

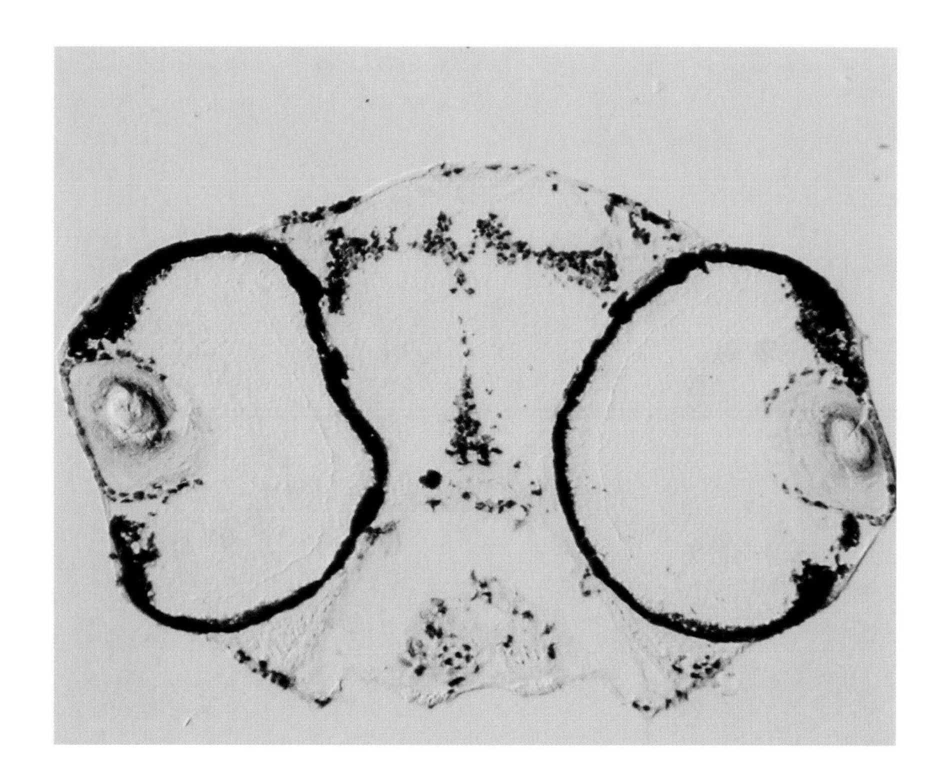

Chapter 3

Analysis

3. Interpretation of Data—How to Use the Atlas

Clearly, the zebrafish represents the primary neurogenetic and neurobiological vertebrate model system regarding accessibility for molecularbiological studies. At the pace such data are presently accumulating, the zebrafish could easily be the first vertebrate organism wherein nearly all gene regulator cascades guiding the development of neuronal (and glial) subtypes within a differentiating neural tissue are functionally described. Especially, the emergence of new tools for investigation, such as high resolution confocal microscopy for analyzing whole-mount (embryonic, larval, and adult) brains, or the establishment of fast increasing numbers of transgenic (e.g., Gong et al. 2001), enhancer trap (e.g., Kawakami et al. 2000) and BAC clone derived (e.g., Yan et al. 1998) zebrafish lines, will make the zebrafish a first choice vertebrate model for decovering mechanisms and gene regulatory networks leading to a fully differentiated brain. The present atlas is envisaged as a prerequisite for such work. In the light of different molecular markers at different stages, the atlas delivers a basic reference of neuroanatomical designations for the postembryonic zebrafish brain that are widely accepted in other vertebrates. The atlas also delivers locally differing distributions of various fundamental cell states during secondary neurogenesis at the level of sectioned material in the developing zebrafish brain, in the same way it is established in other vertebrate model systems. This facilitates the comparison of orthologous expression domains between different species in a neurobiological context and allows eventually to extrapolate hypothetical gene expression data from one species to another (see final comparative chapter).

The general idea is simple: The study of genetic pathways may profit from the degree of definition of the anatomical/developmental paradigm. For instance, the success in decovering molecular mechanisms in *Drosophila*, e.g., the Delta–Notch mediated lateral inhibition leading to the singling out of sensory organ precursor cells, had been based on clear and precise definition of developmental paradigms and anatomical resolution (see Chapter 1). Since the development of the zebrafish brain as a whole is certainly much more complex, decovering molecular mechanisms there relies even more on clearly defined developmental paradigms and anatomical descriptions when it comes to study local differences and specificities in neurogenetic pathways or morphogenetic neurobiological processes (see examples below). Stage-specific neuroanatomical details and neurogenetic contexts, delivered at the level of sectioned material as in this atlas, may additionally help for the interpretation of expression data at the level of wholemount analysis.

This chapter gives an impression of the general developmental dynamics of neurogenesis in the zebrafish brain between 2 and 5 dpf (section 3.1). The diencephalon represents a particularly nice case of postembryonic ventricular neuronal production, a radial glia guided migration and peripheral differentiation. The diencephalon furthermore offers the chance to demonstrate how special local cases of unusual developmental dynamics in the zebrafish brain (i.e., peripheral proliferation and neurogenesis in M1/M2) may be analyzed within this general framework. As a second case, the cerebellum–rhombic lip region will be discussed as a prime example of neurogenesis in the context of extensive tangential migration. We will also show how the bromodeoxyuridine (BrdU) method, as well as the combination and addition of cellular markers, may be used to improve the understanding of such locally differing developmental dynamics of neurogenesis (section 3.2). Finally, we will give an example of how to implement new gene expression data into the existing picture delivered by the atlas (section 3.3).

3.1. General and Local Dynamics of Neurogenesis

A general account on the developmental dynamics of postembryonic proliferation, early neuronal determination and differentiation using molecular markers related to the *neurogenin/neurod* pathway has been given in Chapter 2. In the 2-day zebrafish brain, the general picture is a three-partitioned gray matter with ventricular proliferative (i.e., PCNA/*notch1a* positive) cells, freshly determined (i.e., *neurod* positive) neuronal cells migrating towards the periphery of the gray matter, and finally, differentiating neuronal cells (Hu-positive) at the lateral border of the gray matter. The clear spatial separation of these cell masses in most brain areas likely represents three phases of neuronal development and is a unique feature of the 2-dpf zebrafish brain. Beginning with 3 dpf, Hu-positive cells extend down to the ventricular proliferating cells and are essentially complementary to these proliferative zones (Mueller and Wullimann 2002b). Thus, Hu-proteins are now generally indicative of postmitotic neuronal cells, although we have noted that—as an exception—some clearly postmitotic differentiated structures remain Hu-negative (Mauthner neuron, subcommissural organ; see Chapter 2). Regions maintaining higher levels of proliferation generally show a retarded maturation, as can be judged by a relatively later expression of Hu-positive cell populations in these regions (e.g., dorsal thalamus, optic tectum, and cerebellar plate including rhombic lip). Finally, there are regions with high proliferative activity, but without *neurogenin/neurod* gene expressions (i.e., subpallium, preoptic region, ventral thalamus, hypothalamus); as we shall see later, *Zash1a* is expressed there (see section 3.3).

3.1.1. *The Exception to the Rule: Peripheral Proliferation*

The increasing differentiation during postembryonic zebrafish brain development allows to recognize two migrated areas in the diencephalon at 3 dpf, i.e., the migrated pretectal region M1 (the future superficial pretectum) and the migrated posterior tubercular region M2 (the future preglomerular region). While diencephalic proliferation zones are nicely restricted to ventricular positions and are essentially complementary to Hu-positive cells between 3 and 5 dpf (fig. 8E1, E2 and fig. 9E1, E2), additional, non-ventricular proliferative clusters and sites of neurogenesis are seen in the peripherally migrated M1 and M2 beginning with 3 dpf (PCNA, *neurod*, fig.8A–D and fig. 9A–D; compare also with Chapter 2). Similarly, in the developing cerebellar plate, the non-ventricularly located proliferation zone of the external granular layer (EGL) is visible for the first time at 3 dpf (see below). These three regions build the only obvious exceptions to the rule that proliferation and neurogenesis in the postembryonic zebrafish brain takes place at ventricular sites.

3.1.2. *Unusual Peripheral Proliferation and Neurogenesis in the Diencephalon: M1 and M2*

Corresponding to ongoing relatively strong (ventricular) proliferation around 4–5 days in the pretectal, dorsal thalamic, and ventral posterior tubercular area, expression of *neurod* is upheld in cells adjacent to these diencephalic proliferation zones (figs. 8 and 9), despite a general strong downregulation of *neurod* expression at 5 days in most other places. Additionally, *neurod*-positive clusters are seen in M1 and M2, and they even seem to increase in size with age.

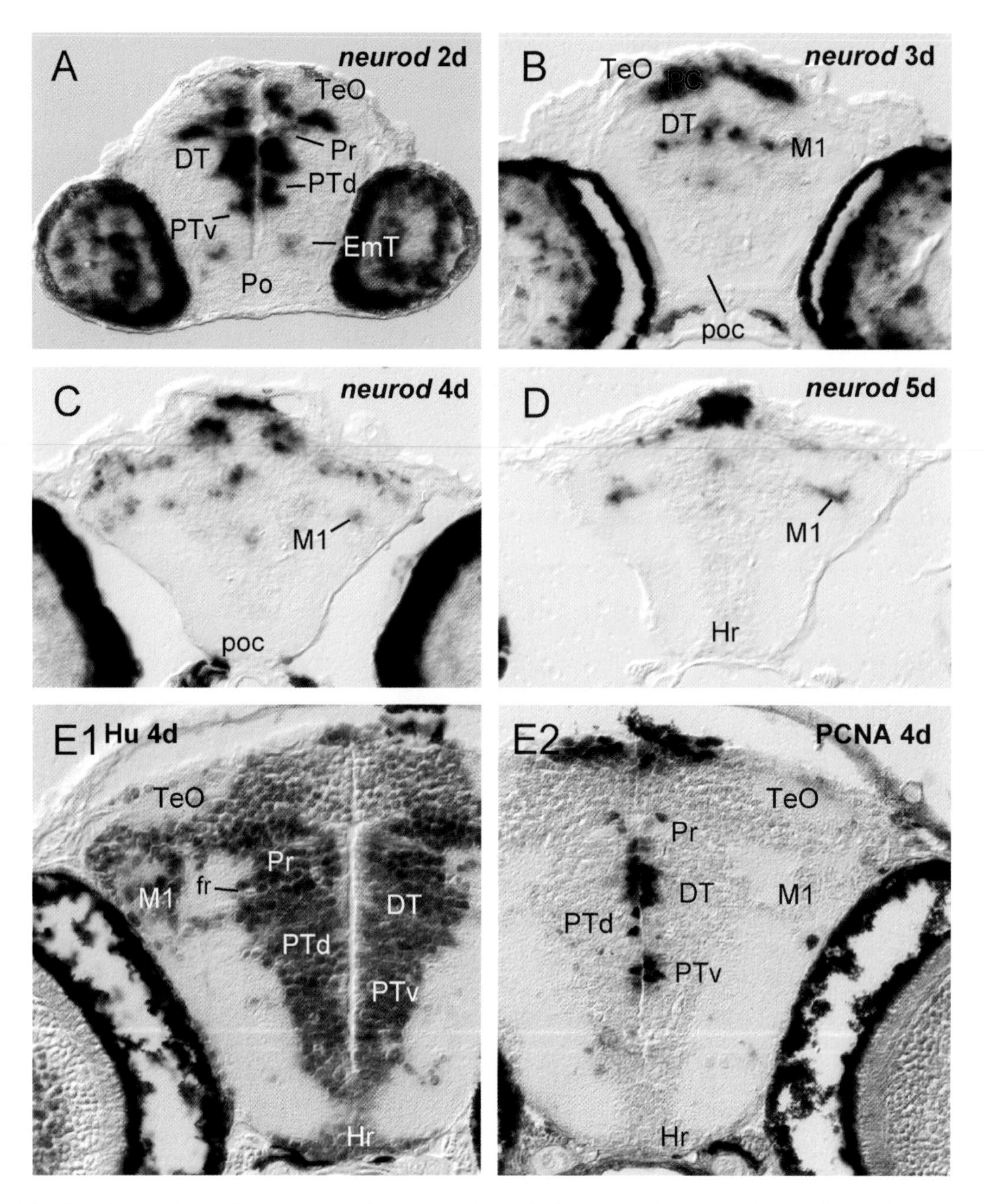

Fig. 8. Neurogenesis in the postembryonic zebrafish diencephalon at the level of the migrated pretectal area M1. (A–D) Expression of *neurod* in the postembryonic zebrafish brain. Corresponding diencephalic sections showing differentiated neuronal cells (E1) and proliferative cells (E2). Abbreviations: DT, dorsal thalamus (thalamus); EmT, eminentia thalami; fr, fasciculus retroflexus; Hr, rostral hypothalamus; M1, migrated pretectal area; Po, preoptic region; poc, postoptic commissure; Pr, pretectum; PTd, dorsal part of posterior tuberculum; PTv, ventral part of posterior tuberculum; TeO, tectum opticum.

Correspondingly, one observes proliferation (PCNA) in M1 and M2 at least up to 10 dpf (unpublished observations). The dorsal thalamic proliferation is one of the strongest and most persistent proliferation zones. Numerous cells apparently stemming from this proliferation zone seem to emanate laterally from it, to express

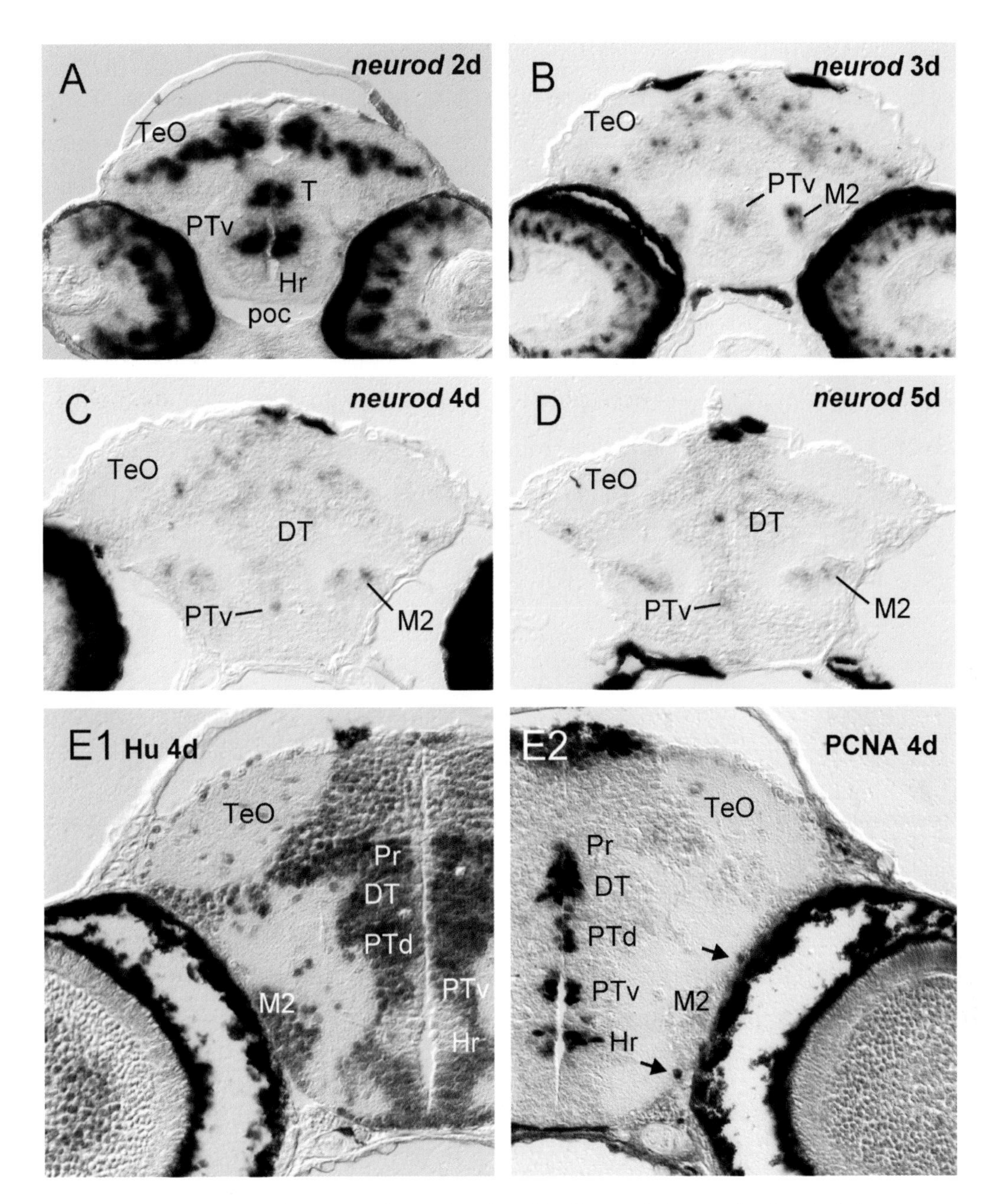

Fig. 9. Neurogenesis in the postembryonic zebrafish diencephalon at the level of the migrated posterior tubercular area M2. (A–D) Expression of *neurod* in the postembryonic zebrafish brain. Corresponding diencephalic sections show differentiated neuronal cells (E1) and proliferative cells (E2). Abbreviations: DT, dorsal thalamus (thalamus); Hr, rostral hypothalamus; M2, migrated posterior tubercular area; poc, postoptic commissure; Pr, pretectum; PTd, dorsal part of posterior tuberculum; PTv, ventral part of posterior tuberculum; T, mesencephalic tegmentum; TeO, tectum opticum.

neurod and to migrate into the pretectal region M1 (fig. 8B–D). This *neurod* migratory stream is strongest at 3 dpf, but *neurod* remains strongly expressed in M1 at least up to 5 dpf (fig. 8D). In contrast, *neurod* expression is almost absent at ventricular diencephalic sites at this stage.

Somewhat more caudally, *neurod* is strongly expressed adjacent to ventricularly located proliferation zones of the ventral posterior tubercular area (as well as in mesencephalic regions, such as the optic tectum and midbrain tegmentum; fig. 9A). Within the migrated posterior tubercular region M2, *neurod* expression is strongly upheld at 3 dpf (fig. 9B) and maintained into at least 5 dpf (fig. 9C,D). The dorsal thalamic *neurod* expression is still seen at these caudal levels up to 5 dpf (fig. 9D). As in M1, proliferation (PCNA) occurs in M2 from 3 dpf onwards (shown for 4 dpf; fig. 9E2, arrows; see also fig. 13). In these special peripheral locations M1 and M2 (as well as in the cerebellar EGL), the expression of *neurod* likely defines neuronally determined and possibly also still proliferating cells (Mueller and Wullimann 2002b), as has definitely been shown for the mammalian *neurod* ortholog *neuroD/BETA2* in the EGL of the mouse (Lee et al. 2000).

3.1.3. The Cerebellar and Rhombic Lip Region: Pervasive Tangential Migration

In the general region of the zebrafish cerebellum (i.e., valvula, cerebellar plate, lateral hinge point between cerebellar plate and medulla oblongata, eminentia granularis, and rhombic lip), a different situation exists. Maturation is also delayed here, as there is an early strong PCNA/*neurod* positivity (with the valvula being even more delayed than the other structures mentioned, see Chapter 2), and a corresponding absence of differentiated (Hu) cells (fig. 10A1–A3,B). There is a massive *neurod* expression extending throughout the cerebellar plate and extending lateroventrally from the rhombic lip into the medulla oblongata (fig. 10B). The latter *neurod* expression in the lateral medulla oblongata is seemingly more lateroventrally migrated between 3 and 5 dpf (fig. 10B–D; compare with Chapter 2). This lateroventral displacement of medullary *neurod* expression is clearly consistent with the general knowledge about tangential migration from the rhombic lip into the base of the medulla oblongata in vertebrates (fig. 11; see also Chapter 1). It is also consistent with the increasing number of differentiated structures arising in the ventral zebrafish medulla oblongata between 3 and 5 days that are known to arise from the rhombic lip in amniotes, such as the inferior olive and the lateral reticular nucleus (compare with Chapter 2).

The zebrafish cerebellar plate is a rather late developing structure where strong proliferation is seen until at least 5 dpf, in contrast to the medulla oblongata

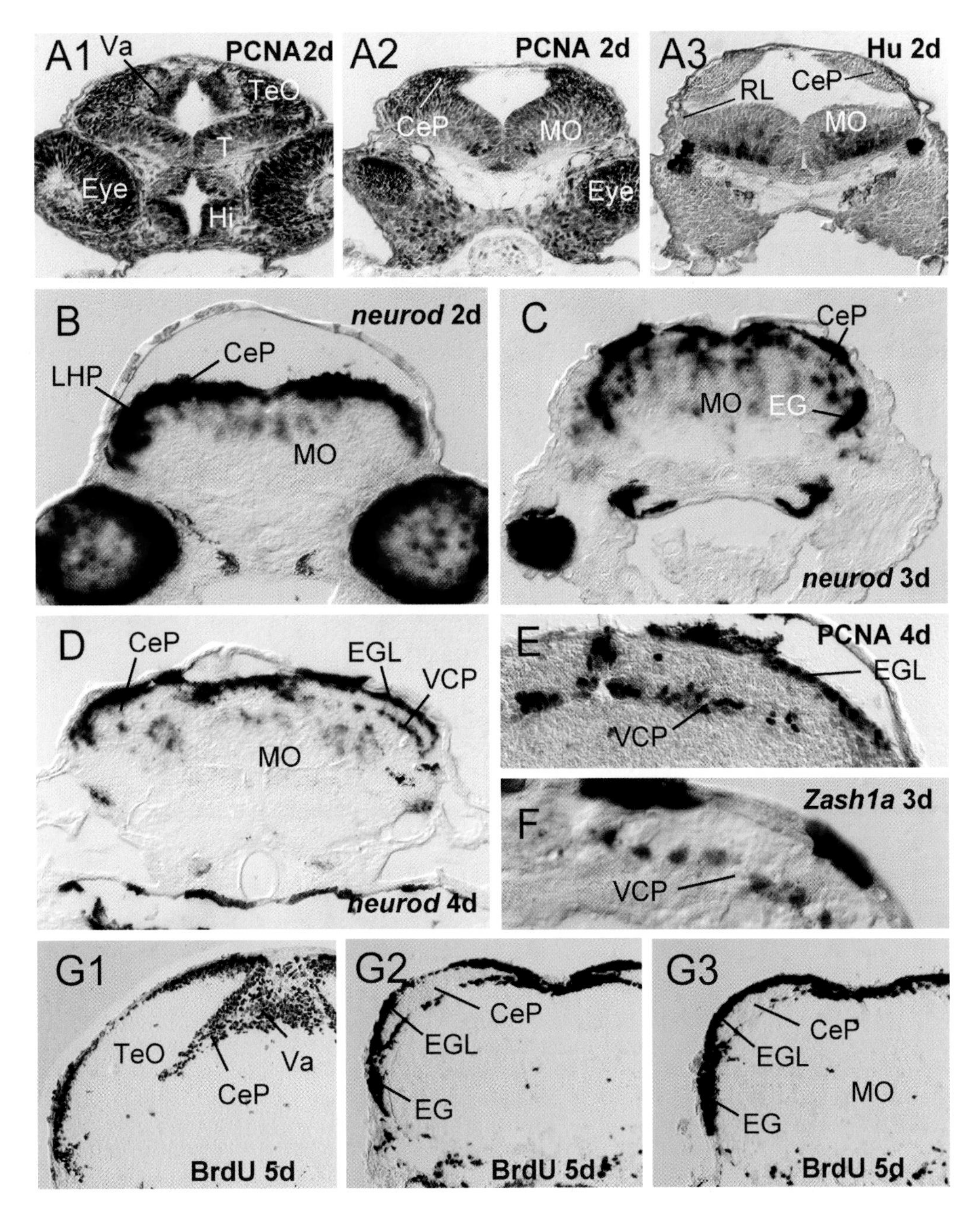

Fig. 10. Many proliferative cells (PCNA; A1; A2), but absence of differentiated cells (A3) characterize the early postembryonic cerebellum. (B–D) Expression of *neurod* in the postembryonic zebrafish brain. (E) Proliferation in the external granular and ventral cerebellar proliferative layer. (F) Expression of *Zash1a* in the ventral cerebellar proliferative layer. (G1–G3) Proliferation in the zebrafish cerebellum demonstrated with BrdU. Abbreviations: CeP, cerebellar plate; EG, eminentia granularis; EGL, external granular layer; Hi, intermediate hypothalamus; LHP, lateral hinge point; MO, medulla oblongata; RL, rhombic lip; T, mesencephalic tegmentum; TeO, tectum opticum; Va, valvula cerebelli; VCP, ventral cerebellar proliferative layer.

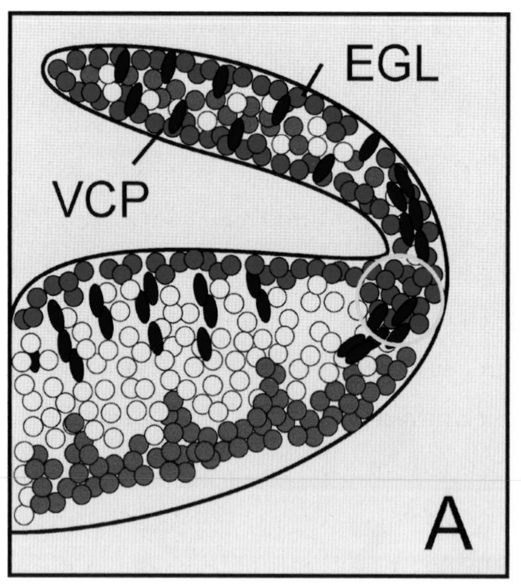

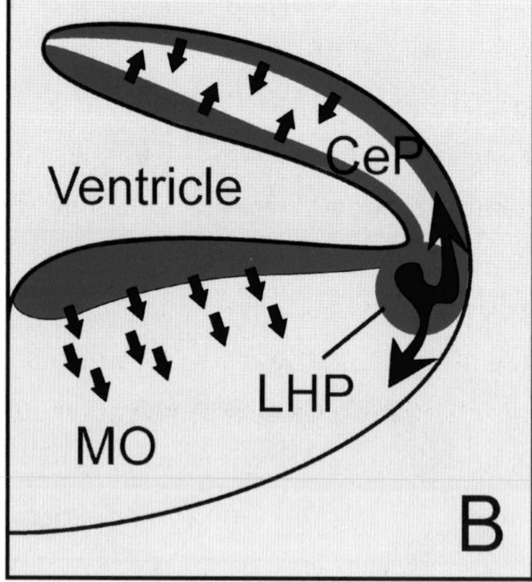

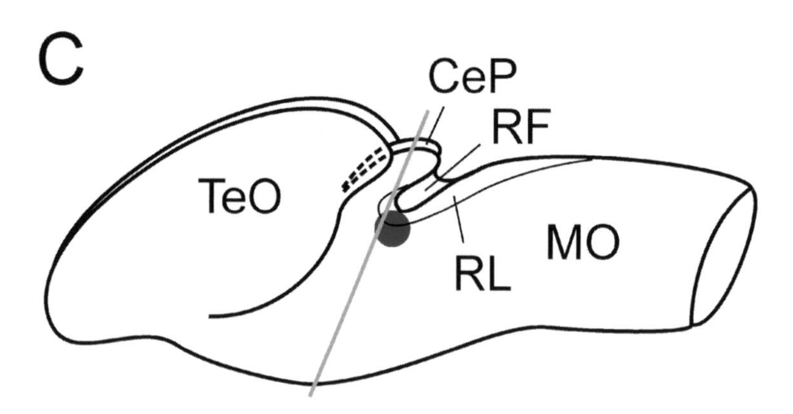

Fig. 11. Tangential migration during neurogenesis in the postembryonic zebrafish cerebellum and rhombic lip region. (A) Schematic distribution of proliferative (green), migrating (blue) and differentiating (red) cells in the postembryonic zebrafish cerebellar plate and medulla oblongata. In addition to radial migration within both medulla oblongata and cerebellar plate (short arrows in B), there is additional tangential migration originating in the rhombic lip into the EGL and into the peripheral medulla oblongata (long double arrow). (C) Lateral schematic view of zebrafish midbrain and hindbrain indicating level of sections (A) and (B). Abbreviations: CeP, cerebellar plate; EGL, external granular layer; LHP, lateral hinge point; MO, medulla oblongata; RF, rhombic fossa; RL, rhombic lip; T, mesencephalic tegmentum; TeO, tectum opticum; VCP, ventral cerebellar proliferative layer.

(see Chapter 2). At 2 dpf, proliferation is very strong in the cerebellar plate (fig. 10A2). Accordingly, no differentiating Hu-expressing cells can be detected (fig. 10A3). Expression of *neurod* is also very strong at 2 dpf where essentially all cerebellar plate cells seem to express this gene (fig. 10B), apparently including many, if not all, proliferative cells. Starting with 3 dpf, a second, peripheral proliferative zone emerges and, thus, separates the cerebellar plate into two proliferative layers, the ventral proliferative layer and the EGL (shown nicely with

PCNA at 4 dpf; fig. 10E; and with BrdU at 5 dpf; fig. 10G2,G3). Correlated expression of *neurod* is found in cells in or near the EGL and the ventral proliferative layer of the cerebellar plate (VCP) (best seen at 4 dpf in fig. 10D).

These findings are consistent with the knowledge in amniotes that the rhombic lip generates cells which migrate tangentially into the utmost periphery of the developing cerebellar plate to form a proliferative EGL there. The latter, in turn, produces the cerebellar granular cells, which migrate (radially) down to the base of the cerebellum (see Chapter 1). Thus, it is very likely that also the EGL in the zebrafish represents a case of tangential migration. Indeed, Köster and Fraser (2001) have directly demonstrated the ontogenetic origin of the EGL cells from the rhombic lip by showing migration of early cerebellar cells in a live imaging study in the zebrafish (as well as even more upper rhombic lip cells that migrate ventrally and populate extracerebellar, medullary territories, confirming also tangential migration into the medulla oblongata discussed above).

The mammalian *neurod* ortholog *neuroD/BETA2* is expressed in the cerebellar inner and EGL (in the latter in proliferative cells; Lee et al. 2000; see above) similar to the situation in the postembryonic zebrafish described above. In addition, a second (*Drosophila atonal*-related) bHLH factor, *MATH-1*, plays a crucial early role in the mammalian rhombic lip and its descendant cells and is later exclusively expressed in the mammalian EGL (not in other cerebellar regions; reviewed in Hatten et al. (1997), Wang and Zoghbi (2001)). A corresponding gene would be highly desirable to be identified in the zebrafish for further studying tangential migration in the rhombic lip region, including its role in zebrafish granular (i.e., glutamatergic) cell development. Equally interesting is that we found the (bHLH) *Zash1a* gene expression restricted or close to the ventricular proliferative layer (and not in the EGL; fig. 10F). As we have hypothesized before (Wullimann and Mueller 2002), this could be related to the production of GABAergic cerebellar cells, such as the Purkinje cells.

3.2. The BrdU Approach

Despite the general decrease of proliferation (PCNA) and increase in differentiation (Hu) in the zebrafish brain between 2 and 5 days described above, the PCNA assay also suggests a contra-intuitive increasing peripheral proliferation in the migrated pretectal area M1, in the migrated posterior tubercular area

M2, and in the cerebellar EGL, as discussed above. Similarly, a continuing elevated *neurod* expression in these places—despite a massive downregulation of *neurod* in most other brain areas—reveals a related ongoing neuronal determination there. These observations of local differences in developmental dynamics raise some problems regarding the corroboration of proliferation and neurogenesis in these peripheral regions.

The PCNA assay has some disadvantages for detecting proliferative cells. Its rather harsh demasking method (i.e., microwave cooking) prior to immunolabeling compromises the subsequent use of additional antibodies or a combination with in situ hybridization. Moreover, PCNA might be detectable in cells for a while after mitosis and is, therefore, not fully reliable for determining exclusively mitotic cells (discussed in Mueller and Wullimann (2002b)). We have, therefore, turned to the more reliable BrdU method to determine proliferative cells. In our hands, this method has been very fruitful—in combination with other markers—for determining the neurogenetic state of particular cell populations during neurogenesis (e.g., for showing truly mitotic cells), but it also offered a variety of additional opportunities to address certain developmental problems which cannot be addressed with the PCNA method. To explain our experiments using differing ways of BrdU labeling, it is worthwhile to give some theoretical details (summarized in fig. 12), as we believe that this methodology is of general value for many studies in the zebrafish.

Animals exposed to BrdU (5-*br*omo-2′-*d*eoxy*u*ridine) incorporate this thymidine analog into the DNA of mitotic cells in the S-phase. The zebrafish—in contrast, for example, to the mouse—offers the advantage for external BrdU exposure (although there is naturally a retardation in uptake compared to experiments where BrdU is injected into the organism or applied in cell culture systems because of reduced surface-to-volume ratio and impermeability of the epidermis). The exposure time of externally applied BrdU to zebrafish embryos and larvae is stage and purpose dependent, as with increasing age, individual proliferation activity differences arise because of decreasing synchronization of development.

Our incubation of zebrafish larvae (4 dpf) started around midnight and they were sacrificed sequentially during the following day. Increasing incubation times led to the distinction of four BrdU approaches which can be used for different purposes: the BrdU pulse label selectively detects S-phase cells, the BrdU saturation label visualizes all proliferative cells, the BrdU long-term label gives

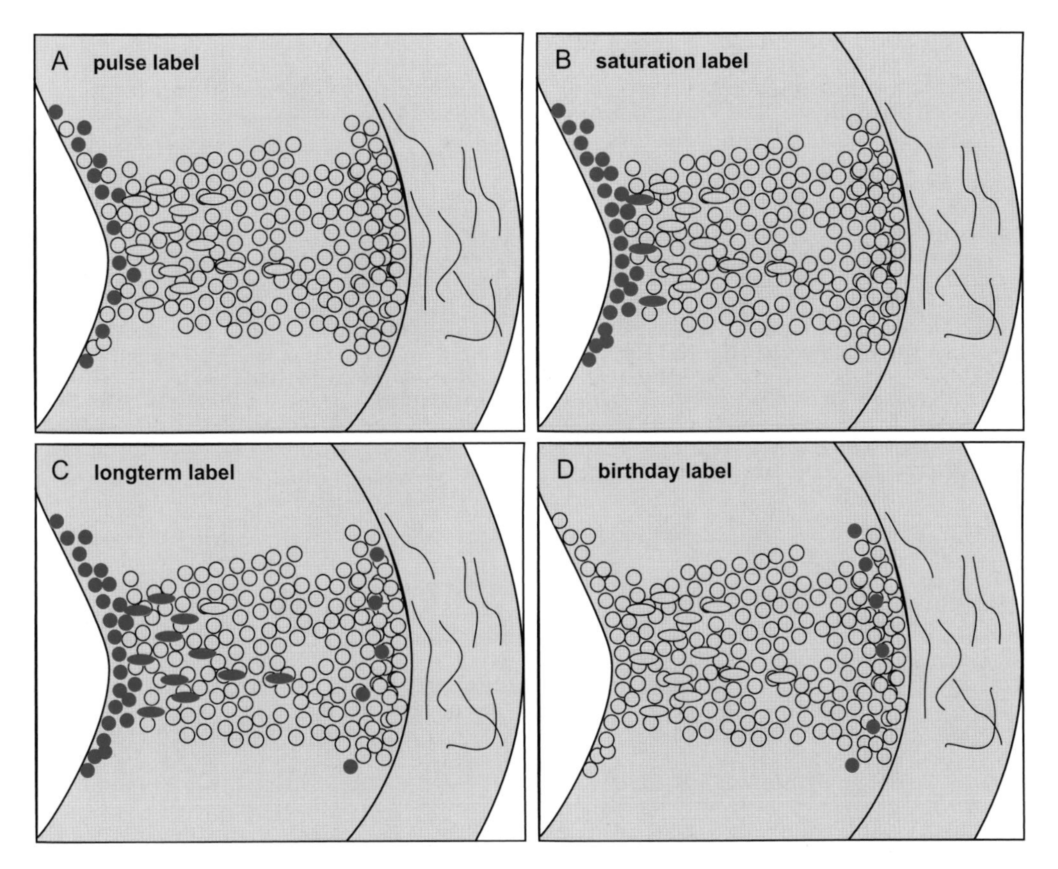

Fig. 12. Schematics show labeled BrdU cells (green) after differing incubation paradigms. (A) Pulse label. (B) Saturation label. (C) Long-term label. (D) Birthday label. For explanations, see text.

a non-quantitatively estimate of the relation between mitotic and newborn migrating cells, and the BrdU birthday label demonstrates cell fates of cells born during the distinct time window of development when BrdU has been applied (see also Mueller and Wullimann 2002b) and these approaches shall be defined shortly in the following.

3.2.1. BrdU Pulse Label

First mitotic cells in the zebrafish brain could be detected after an incubation time of 8–10 h for 5 dpf zebrafish larvae. This method detects only S-phase cells. The incubation time in this approach should be as short as possible and the animal has to be sacrificed immediately after incubation. Compared with a PCNA preparation, which shows all proliferative cells in all proliferative zones, a BrdU pulse label reveals only a subpopulation of PCNA-positive cells (fig. 12A) scattered within a given proliferative zone. No BrdU positive cells are detectable in a migrated position (despite some in those peripheral proliferative zones seen with PCNA mentioned above, i.e., M1, M2, cerebellar EGL).

3.2.2. *BrdU Saturation Label*

A BrdU incubation time of 12–16 h revealed what we termed *BrdU saturation label* in the zebrafish brain (fig. 12B; Mueller and Wullimann 2002b). Our experiments allowed to define the saturation label approach using three features. (1) The BrdU saturation label approach visualizes practically every proliferative cell and the resulting overall picture is qualitatively similar to that delivered by other mitotic markers (such as PCNA; Wullimann and Puelles 1999; Wullimann and Knipp 2000). (2) At the same time, it avoids massive labeling of BrdU-positive, migrated postmitotic cells. The few, freshly postmitotic cells that are BrdU labeled can be differentiated from those in the proliferative zones because of their slightly migrated position, sometimes because of their weaker BrdU staining and eventually because of their cell shape (somewhat flattened appearance compared to round shape and large nuclei of proliferative cells; fig. 13D3). The BrdU saturation label can be used in conjunction with markers for neuronally determined cells (such as *neurod* transcripts; or earliest phenotypically differentiating neuronal cells, e.g., Hu-proteins). This allows to shed light on the developmental phase of neurogenesis a cell is in. For instance, if *neurod* is expressed in freshly determined neuronal cells, some BrdU and *neurod* stained cells may be expected immediately lateral to the strongly and entirely labeled ventricular proliferation zones. (3) Importantly, such a saturation label must show an absence of long-distance migrated postmitotic cells (except again in peripheral proliferative zones seen with PCNA mentioned above, i.e., M1, M2, cerebellar EGL). The length of the cell cycle is rapidly increasing in the early zebrafish CNS; it reaches 4 h already during gastrulation at 10 hpf (Kimmel et al. 1994) and at 24–28 hpf the cell cycle length amounts to 10 h (Li et al. 2000). With a BrdU incubation time of 12–16 h used in the 4–5 dpf zebrafish brain for reaching a saturation label, we believe that this incubation time likely affects one (or at the most two) cell cycles.

3.2.3. *BrdU Long-Term Label*

We used different long-term labels (18, 24, 30 h) as controls to saturation labels because the increasing incubation times should reveal a consistent

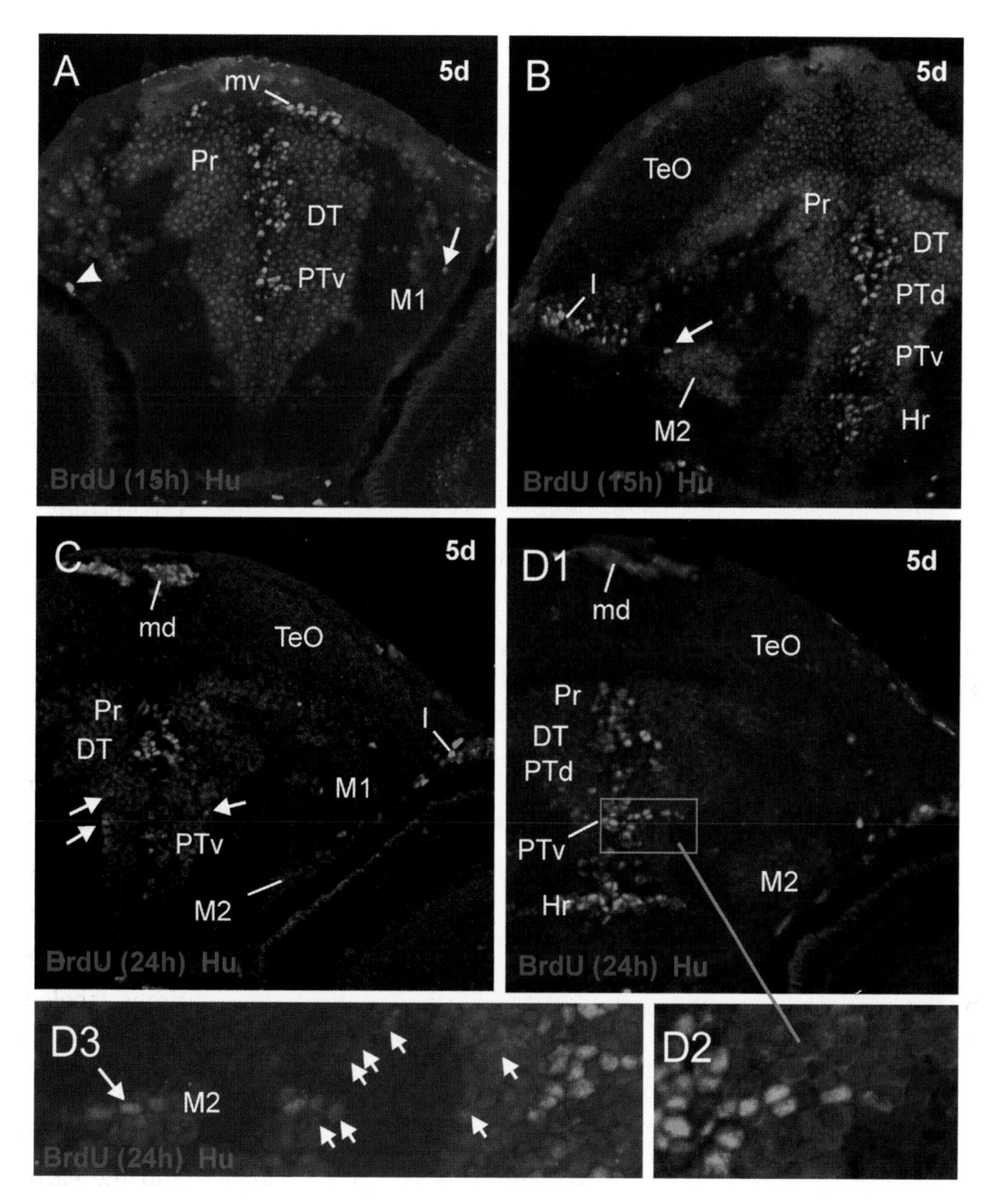

Fig. 13. Neurogenesis in the postembryonic zebrafish diencephalon. Saturation label at the level of the migrated pretectal area M1 (A) and migrated posterior tubercular area M2 (B). Arrows in (A)–(B) designate peripheral proliferating (BrdU labeled) cells (green). Arrowhead in (A) points to BrdU labeled meningeal cell. Long-term label at the same diencephalic levels (C)–(D). Arrows in (C) designate migrated postmitotic (BrdU labeled) cells. (D1)–(D2) Migration streams are visualized with BrdU in long-term label. (D2) Double-labeled cells for Hu (red) and BrdU (green) confirm neuronal identity of proliferative cells in M2. Modified after Mueller and Wullimann (2002b). See text for more details. Abbreviations: DT, dorsal thalamus (thalamus); Hr, rostral hypothalamus; M1, migrated pretectal area; M2, migrated posterior tubercular area; md, mediodorsal tectal proliferation; mv, medioventral tectal proliferation; poc, postoptic commissure; Pr, pretectum; PTd, dorsal part of posterior tuberculum; PTv, ventral part of posterior tuberculum; T, mesencephalic tegmentum; TeO, tectum opticum.

increase of BrdU-positive cells in increasingly more peripheral positions. A BrdU long-term label marks substantially more cells in addition to proliferative ones (fig. 12C) and these more peripheral postmitotic cells often display an ovoid cell shape typical for migrating cells. Depending on the incubation time, they may have reached their final positions. BrdU long-term labels are somehow comparable with pictures and data delivered by sections of green fluorescent protein (GFP) transgenic lines. Transgenic zebrafish lines of neurogenic or proneural genes are of immense interest for fate studies, because the expression of the reporter protein is longer upheld than the expression of the gene of interest (mRNA as well as the protein). Therefore, similar to a comparison of a BrdU saturation label to a long-term label, the comparison of in situ hybridized sections for a given gene expression compared to corresponding sections of a transgenic GFP-line of the gene of interest may reveal the fates of the descending cell lines depending on the gene of interest. Since proneural or neurogenic genes are sometimes regionally specifically expressed, descending cells could be easily studied in this way regarding their local destinations and neuronal fates (e.g., the resulting transmitter type). Compared with additional, BrdU incubated brains at that particular stage, important insights into the spatiotemporal dynamics and significances of genes involved in certain neurogenetic steps would emerge.

3.2.4. BrdU Birthday Label

In the birthday label approach, the survival time after incubation is very long (weeks to months). It allows to potentially determine the adult location of cells, which were given off by proliferative neural cells which went through their last mitosis (i.e., S-phase) during incubation. During this survival time, the BrdU-labeled cells may migrate over long distances and reach the locations of their final destination (fig. 12D; see for example Schmidt and Roth (1993)).

3.2.5. Analysis of Diencephalic Migrated Sites of Neurogenesis

Using the BrdU saturation label in the zebrafish, mitotic activity in peripheral regions could be demonstrated, i.e., in the migrated pretectal region M1 and migrated posterior tubercular region M2 (fig. 13A,B; arrow: peripheral BrdU cell; arrowhead designates BrdU-positive meningeal cell) and in the cerebellar

EGL (fig. 10G2,G3). As expected, there were no BrdU cells in between the ventricular proliferation zones and these peripheral regions. The analysis of the cerebellar plate has already been discussed above and, thus, we will focus here on the diencephalon (M1/M2).

In Hu-protein/BrdU double immunostains, some of these peripheral diencephalic proliferative cells in M1 and M2 revealed their neuronal nature, which is in line with the continued *neurod* expressing clusters there (see above). Also as expected, long-term BrdU label revealed increasing numbers of additional, migrating BrdU labeled cells outside the proliferation zones (arrows in fig. 13C). Furthermore, some particularly strong streams of apparently migrating, BrdU-positive cells could be demonstrated, e.g., one originating in the ventral posterior tubercular proliferation zone (fig. 13D1–D3) and one migrating stream originating in dorsal thalamic/pretectal proliferation zones (not shown). It was also possible to show that only ventricular and peripheral BrdU-positive cells were at the same time Hu positive (fig. 13D3; long arrow), but not the BrdU-positive migrating cells (fig. 13D3; short arrows).

3.3. How to Fit in New Gene Expressions: *Zash1a/Zash1b*

In Chapter 2, we have treated some gene expressions and additional genetic markers involved in zebrafish brain neurogenesis in the *neurogenin/neurod* pathway. As mentioned there, additional bHLH genes are involved in neurogenesis and we will look here—as an example—at how the *achaete scute* zebrafish orthologs fit into the picture, i.e., *Zash1a* (*ascl1a*) and *Zash1b* (*ascl1b*).

Several *Drosophila achaete-scute* orthologs have been identified in vertebrates. One of them, *Mash1* (*mammalian achaete-scute homolog 1*), is involved in the specification of neuronal subtype identity, e.g., in determination of noradrenergic and GABAergic neuronal cells in the CNS (Lo et al. 1991, 1998; Hirsch et al. 1998; Casarosa et al. 1999; Fode et al. 2000; He et al. 2001; Parras et al. 2002; Schuurmans and Guillemot 2002). In murine species, the functional roles of *Mash1* are reflected by complex region- and stage-dependent expression patterns which are complementary to those of *neurogenin1* in the spinal cord and forebrain (Guillemot and Joyner 1993; Sommer et al. 1996; Ma et al. 1997; Horton et al. 1999; Torii et al. 1999; Fode et al. 2000) at a particular developmental stage (mouse: E12.5–13.5, rat: E14.5).

3.3.1. *How Is the Situation in the Zebrafish?*

Two *achaete-scute* orthologs, *Zash1a* and *Zash1b*, have been characterized during embryonic CNS development (Allende and Weinberg 1994). Both genes are strongly expressed in the embryonic zebrafish brain and spinal cord, showing sharply delineated expression domains reaching from ventricular to pial surface, and their expression patterns are completely different from each other during embryogenesis (up to 30 hpf). In any case, the two genes are not expressed in the zebrafish embryonic brain in a fashion comparable to *Mash1* in the mouse and rat (Allende and Weinberg 1994) and a function in the early patterning of the zebrafish brain has been suggested based on the expression patterns of *Zash1a* and *Zash1b* (Allende and Weinberg 1994).

Thus, we have recently studied the later expression of *Zash1a* (Wullimann and Mueller 2002) and *Zash1b* (Mueller and Wullimann 2003) and have found that the expression domains of both *achaete-scute* orthologs drastically changes in the postembryonic brain compared with embryonic zebrafish stages (compare with Allende and Weinberg 1994). *Zash1b* is expressed in all proliferation zones in the 2 dpf zebrafish brain, except for the retinal peripheral edge. With the latter exception, the general brain expression patterns are very similar to those of *notch1a*, which is documented in detail in Chapter 2 and *Zash1b* will, therefore, not be further elaborated on here. Nevertheless, our report on *Zash1b* gene expression (Mueller and Wullimann 2003) has been the first description of an *achaete-scute* ortholog to be exclusively expressed in proliferative CNS cells of any vertebrate species so far and possibly allows to improve the comparability of vertebrate CNS proliferative zones with proneural clusters defined in *Drosophila*.

Also *Zash1a* expression domains in the postembryonic zebrafish brain (i.e., at 3 dpf) are strongly altered in comparison to the embryonic brain (Wullimann and Mueller 2002). The *Zash1a* brain expression domains are more restricted than those of *Zash1b*. In the following, we will focus on *Zash1a* expression patterns (fig. 14) in the forebrain where such domains are found in phenotypic brain regions homologous to those of the mouse where *Mash1* is expressed, that is subpallium, ventral thalamus, preoptic region, and hypothalamus (i.e., ventral inferior lobe).

Furthermore, as we have seen in Chapter 2, these four regions in question (subpallium, preoptic region, ventral thalamus, and hypothalamus) are characterized by the absence of *neurod* expression (fig. 14). Similar to tectum and cerebellar plate (see Chapter 2), these four regions show delayed maturation as

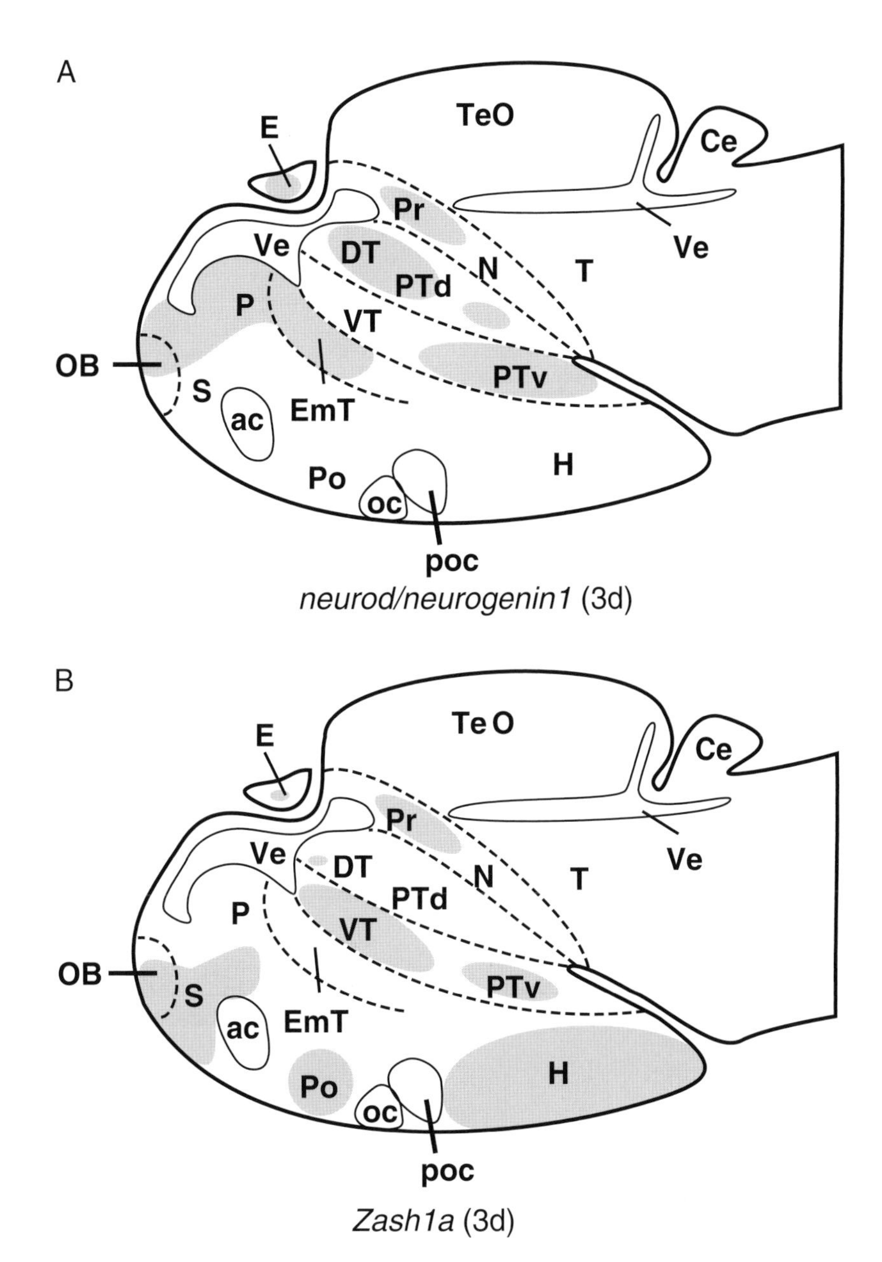

Fig. 14. Early zebrafish forebrain bHLH gene expression shown in lateral views. (A) *neurogenin1* and *neurod*. (B) *Zash1a*. Modified after Wullimann and Mueller (2004b). Abbreviations: ac, anterior commissure; Ce, cerebellum; DT, dorsal thalamus (thalamus); E, epiphysis; EmT, eminentia thalami; H, hypothalamus; Ha, habenula; MO, medulla oblongata; N, region of the nucleus of medial longitudinal fascicle; OB, olfactory bulb; oc, optic chiasma, P, pallium; Po, preoptic region; poc, postoptic commissure; Pr, pretectum; PTd, dorsal part of posterior tuberculum; PTv, ventral part of posterior tuberculum; S, subpallium; T, midbrain tegmentum; TeO, tectum opticum; Va, valvula cerebelli; Ve, ventricle; VT, ventral thalamus (prethalamus).

revealed by a very strong PCNA-immunoreactivity and by a very limited extent of Hu-positive cells. However, absence of *neurod* expression in these regions is not due to an even more immature and proliferative state compared with optic tectum and cerebellar plate, because *neurod* expression is never expressed in these four locations up to 5 days. By then, *neurod* is already largely downregulated in the entire zebrafish brain (see above) and substantial amounts of differentiating Hu-positive neuronal cells can be detected in these four regions as in all brain regions starting with 3 dpf. This suggests that—both in *Zash1a* or *neurod* expressing regions—differentiating neurons exist essentially in the entire zebrafish forebrain already at 3 dpf. Furthermore, expression of *neurogenin1* (*neurog1*) at 2 dpf in the zebrafish brain parallels that of *neurod* (Mueller and Wullimann 2003) in that it too is absent in the subpallium, preoptic region, ventral thalamus, and hypothalamus. Also, in contrast to extensive *neurod* expression in most brain areas, three recently described additional *neurod*-related transcription factors (Liao et al. 1999) revealed very limited expression domains (i.e., *ndr1a* in olfactory epithelium and bulbs, *ndr1b* in olfactory epithelium, *ndr2* only in very early embryo) and, thus, do not replace functionally *neurod* in the four brain regions mentioned.

In summary, this altogether indicates that two well established alternative genetic pathways (i.e., one using *neurog1/neurod*, the other one using *Mash1;* reviewed in Schuurmans and Guillemot (2002)) involved in neurogenesis in the amniote (mammalian) brain are present in homologous phenotypic locations in the anamniote (zebrafish, 3 dpf) brain and that they possibly act similarly in the generation of different neuronal phenotypes, at least in the telencephalon, i.e., subpallial GABAergic interneurons versus pallial glutamatergic projection neurons, (Casarosa et al. 1999; Parnavelas 2000; Schurmaans and Guillemot 2002). The data discussed here furthermore identify the 3 dpf zebrafish brain as the corresponding developmental time point of complementarity of expression domains of the orthologous genes (i.e., *neurod* identified by Blader et al. 1997; *Zash1a* identified by Allende and Weinberg 1994).

We will now focus on the zebrafish diencephalon at 3 dpf to demonstrate how the markers described in Chapter 2 may be used for establishing a zebrafish brain molecular neuroanatomy of the same high resolution already commonly used in amniotes (especially in the mouse) and how one may implement new gene expressions into the basic picture delivered by the markers shown in atlas form in Chapter 2. If one can rely on complete series of brain sections, comparable levels may be analyzed and reveal brain subdivisions that remain hidden using more general approaches for visualizing gene expressions (such as wholemounts). Comparing proliferative (PCNA) with differentiating neuronal (Hu) cells (see fig. 15A), the complementarity of labeling is immediately obvious. This allowed us, for example, to visualize the extent of the zona limitans intrathalamica

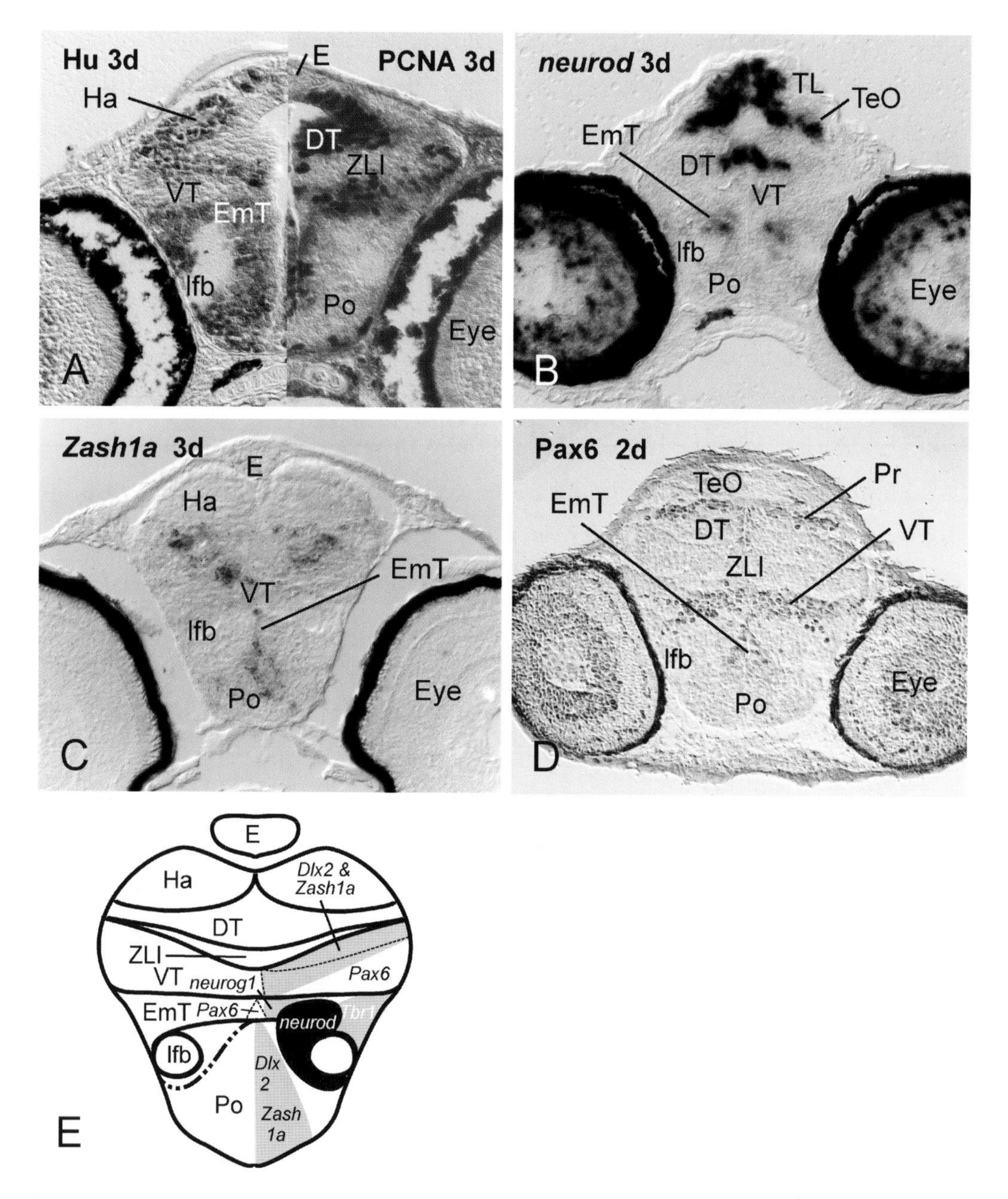

Fig. 15. Neurogenesis in the postembryonic zebrafish diencephalon at the level of the preoptic region/ventral thalamus (prethalamus). (A) Complementarity of proliferative (PCNA; right side) and differentiating (Hu; left side) cells allows basic distinctions of diencephalic subdivisions. Complementarity of *neurod* (B) and *Zash1a* (C) expression in this area. (D) Pax6 protein expression. (E) Interpretative schema including additional gene expressions from the literature (*Dlx2*; Zerucha et al. 2000; *Tbr1*; Mione et al. 2001). See text for more details. Abbreviations: DT, dorsal thalamus (thalamus); E, epiphysis; EmT, eminentia thalami; Ha, habenula; lfb, lateral forebrain bundle; Po, preoptic region; poc, postoptic commissure; Pr, pretectum; PTd, dorsal part of posterior tuberculum; PTv, ventral part of posterior tuberculum; TeO, tectum opticum; TL, torus longitudinalis; VT, ventral thalamus (prethalamus); ZLI, zona limitans intrathalamica.

as the non-proliferative band of Hu-positive cells between dorsal and ventral thalamus. The *neurod* expression at this level illustrates the situation described above: absence of expression in the ventral thalamus and preoptic region, but strong expression in medial optic tectum, dorsal thalamus, and eminentia thalami (fig. 15B). Complementary to *neurod* domains, the *Zash1a* expression is seen at this level in the ventral thalamus and the preoptic region, but not in the dorsal thalamus and not in the eminentia thalami (fig. 15C). We have recently identified the eminentia thalami for the first time in the developing zebrafish (and in any developing teleost, for that matter) demonstrating that it represents a locus of *neurod/neurog1*-guided neurogenesis as in the mouse (Wullimann and Mueller 2004a) and may be discriminated by these expression patterns from the adjacent *Zash1a*-positive preoptic region and ventral thalamus. It is furthermore evident that the Pax6 protein distribution (fig. 15D; for a full account see: Wullimann and Rink 2001) supports these identifications (discussed in Wullimann and Mueller 2004a). If available, well-documented expressions may be included into the analysis from the literature. In our case, that of *Dlx2* (Zerucha et al. 2000) and *Tbr1* (Mione et al. 2001) clearly support the newly revealed identifications of finer diencephalic divisions (fig. 15E). For example, *Dlx2* has seemingly overlapping expression domains in the ventral thalamus and preoptic region with *Zash1a*, while the Pax6 expression of the ventral thalamus is more shifted towards the eminentia thalami. Furthermore, *Tbr1* has a peripheral expression in the newly identified region of the eminentia thalami—similar to the situation in the mouse. We refer to the final chapter for a more thorough comparative discussion of the molecular anatomy of the forebrain of mouse and zebrafish and for more information regarding the mentioned genetic markers taken from the literature. In this paragraph, we wanted to emphasize the power of the methodology presented in this atlas for analyzing the molecular neuroanatomy of the zebrafish brain and, ultimately, for gaining insights into differences between locally acting neurogenetic pathways at a resolution not delivered before in this model system.

Comparison of Fish, Frog, Chick, and Mouse

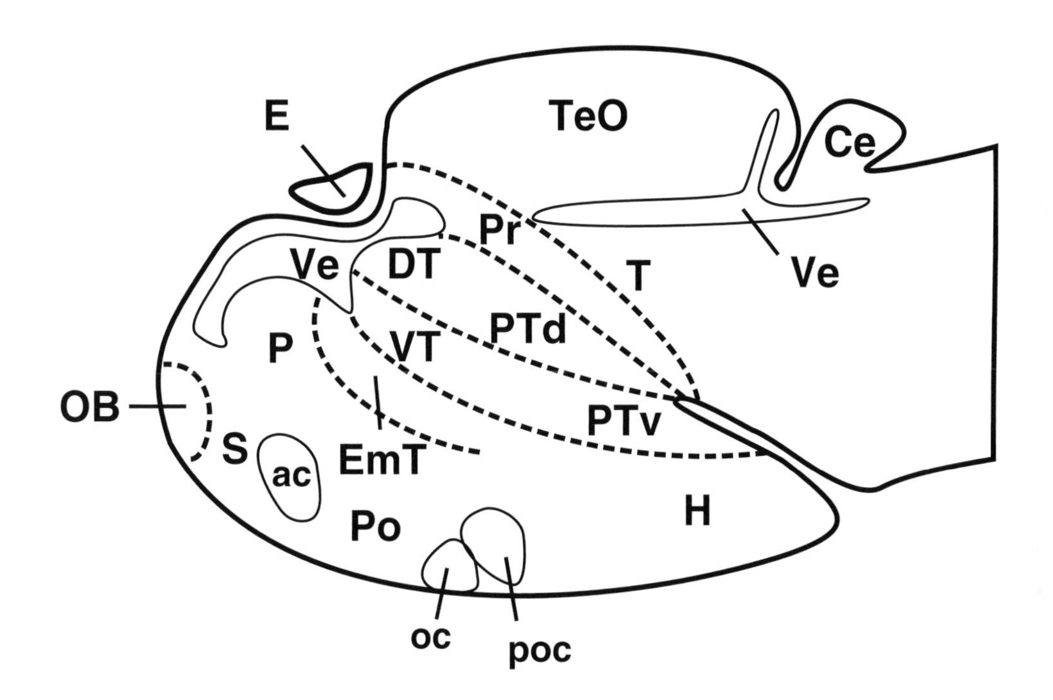

Chapter 4

Model Systems

4. Comparison of Fish, Frog, Chick, and Mouse

In this final chapter, we will compare the zebrafish neurogenetic model system with other vertebrate model animals. Among the latter, the mouse is especially well investigated and the following comparison will, thus, focus on mouse (fig. 16) and zebrafish (fig. 17). Naturally, we will discuss those bHLH genes and the other markers related to neurogenesis that have been treated in this book for the zebrafish, but additional genes will be included into the discussion if information is available from the literature. This comparison of early molecular neuroanatomy will be exemplified for the forebrain.

We have recently reviewed forebrain gene expression data in teleosts (zebrafish) and mammals (mouse/rat), putting it into a more general context of the distribution of early and adult functional (e.g., neurotransmitter-, receptor-, activity related-) forebrain markers, and also including behavioral data related to telencephalic function in teleosts. Importantly, all these data could be consistently interpreted within a newly suggested partial eversion theory of teleostean pallial masses (Wullimann and Mueller 2004b).

4.1. Expression of Basic Helix–Loop–Helix (bHLH) Genes Supports Similar Forebrain Organization in Vertebrates

As discussed at length in Chapter 2, zebrafish brain neurogenic (Delta, Notch) and proneural (bHLH) gene expression patterns between 2 and 5 days reveal locally different and stage-specific activity allowing for a detailed comparison with the mouse. For example, *neurog1* and *neurod* transcripts are absent in four regions of the zebrafish brain (subpallium, hypothalamus, ventral thalamus, preoptic region; figs. 14 and 17). In contrast to what has been described for the embryonic expression of the zebrafish *achaete-scute* homolog *Zash1a* (24 h; Allende and Weinberg 1994), this gene is postembryonically (3 days) clearly expressed in those four *neurog1/neurod*-negative locations just mentioned (and in additional places not considered here; Wullimann and Mueller 2002). These complementary bHLH gene expression patterns are identical to those of

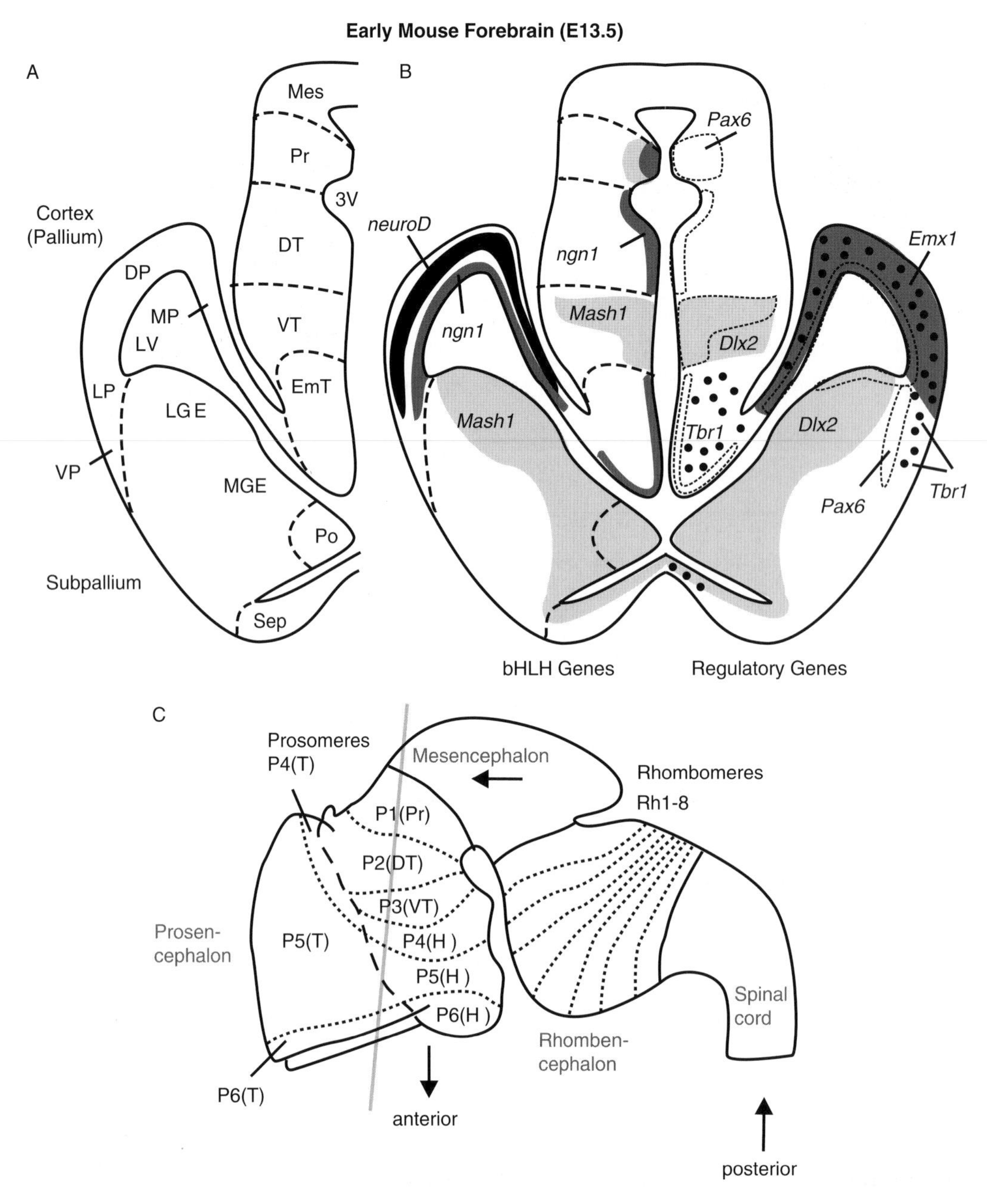

Fig. 16. Early mouse regulatory gene expression patterns (sources for gene expressions, see text; *ngn1* = *neurogenin1*, *nrd* = *neuroD*) shown in transverse section. (A) Neuroanatomical forebrain divisions. (B) Gene expression patterns. (C) Lateral view indicates section level. Modified after Wullimann and Mueller (2004b). Abbreviations: DP, dorsal pallium (isocortex); DT, dorsal thalamus (thalamus); EmT, eminentia thalami; H, hypothalamus; LGE, lateral ganglionic eminence; LV, lateral ventricle; Mes, mesencephalon; LP, lateral (olfactory) pallium; MGE, medial ganglionic eminence; MP, medial pallium (hippocampus); P1–P6, prosomere 1–6; P, pallium; Po, preoptic region; Pr, pretectum; S, subpallium; Sep, septum; T, telencephalon; 3V, diencephalic ventricle; VP, ventral pallium; VT, ventral thalamus (prethalamus).

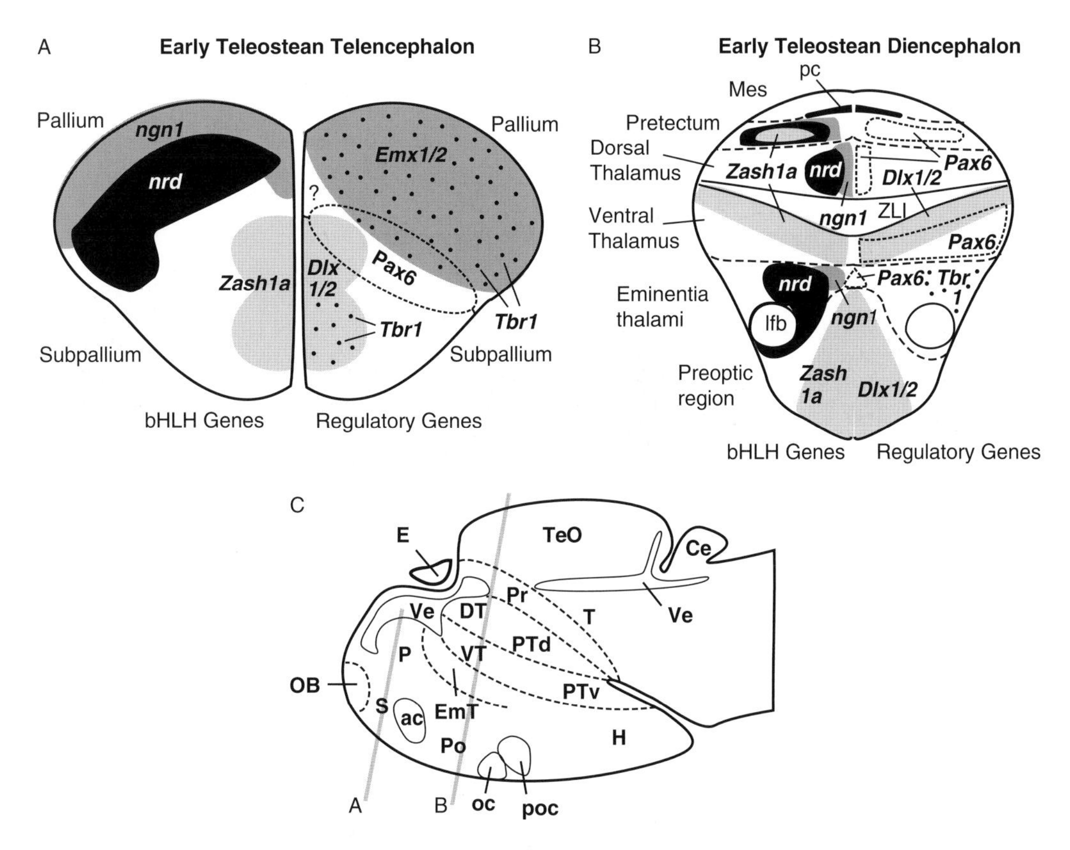

Fig. 17. Early zebrafish regulatory gene expression patterns shown in transverse sections. (A) Characteristic telencephalic gene expressions for pallium (*ngn1* = *neurog1*/*nrd* = *neurod*; after Mueller and Wullimann (2003); *emx1/2*, after Morita et al. (1995), *Tbr1*, after Mione et al. (2001)), subpallium (*Dlx1/2*, after Zerucha et al. (2000); *Zash1a* after Wullimann and Mueller (2002)), and pallial–subpallial boundary (Pax6 protein, after Wullimann and Rink (2001)). Question mark relates to the fact that the exact extent of gene expression in this area needs further clarification, as it is the suspected locus of the ventral pallium. (B) Diencephalic gene expression patterns (citations for genes as above). (C) Lateral view showing section levels. Modified after Wullimann and Mueller (2004b). Abbreviations: ac, anterior commissure; Ce, cerebellum; DT, dorsal thalamus (thalamus); E, epiphysis; EmT, eminentia thalami; Ha, habenula; lfb, lateral forebrain bundle; OB, olfactory bulb; oc, optic chiasma; P, pallium; pc, posterior commissure; Po, preoptic region; poc, postoptic commissure; Pr, pretectum; PTd, dorsal part of posterior tuberculum; PTv, ventral part of posterior tuberculum; S, subpallium; T, mesencephalic tegmentum; TeO, tectum opticum; Ve, ventricle; VT, ventral thalamus (prethalamus); ZLI, zona limitans intrathalamica.

the orthologous genes in murine mammals (fig. 16; Lo et al. 1991, Ma et al. 1997; Horton et al. 1999; Torii et al. 1999; Fode et al. 2000), where the same four areas in the late embryonic (i.e., mouse: E12.5, rat: E14.5) forebrain, i.e., subpallium (striatum), preoptic region, ventral thalamus, and hypothalamus (not shown), have been described to express the mammalian *achaete-scute* homolog *Mash1* in a complementary fashion to *neurogenin1* and *neuroD* (fig. 16B; unfortunately, the situation for *neuroD* is not resolved satisfactorily for the diencephalon in mammals). These expression patterns support the notion that two alternative genetic pathways involved in neurogenesis in the amniote (mammalian) brain are present in homologous phenotypic locations in the anamniote (zebrafish) brain also. Moreover, they are likely related to the generation of subtype- (i.e., transmitter-) specific neuronal cells, since *Mash1* is involved in GABAergic cell production in the mammalian subpallium (see below) and *neurogenin/NeuroD* in glutamatergic cortical cell generation (Casarosa et al. 1999; Parnavelas 2000; Schuurmans and Guillemot 2002; Ross et al. 2003).

4.2. Additional Early Genes Distinguish Pallium and Subpallium in Vertebrates

The amniote telencephalon is characterized by a suite of early active genes that are exclusively expressed either in the pallium (e.g., *Emx1*, *Tbr1*, *reelin*, *Pax6*; note that *Pax6* also extends into the lateral ganglionic eminence) or in the subpallium (e.g., *Dlx1/2*, *Nkx2.1*; Puelles et al. 1999, 2000; note that *Tbr1* has a restricted, additional septal expression and also one in the eminentia thalami). Naturally, these pallial genes in mammals are involved with cortical development whose morphogenesis has been shortly described in Chapter 1. The *reelin* gene is expressed in the earliest postmitotic, peripherally migrating cells of the marginal zone (Cajal-Retzius cells) and codes for the secreted protein Reelin. It has been hypothesized to provide a stop signal (review: Bar et al. 2000) or, alternatively, a detachment signal (extrapolated from studies on—non-radial glia mediated—chain migration in the rostral migratory stream to the olfactory bulb; Hack et al. 2002) for later radially migrating neurons pivotal for correct cortical lamination. The remaining pallial genes discussed code for transcription factors. The T-box gene family (its first identified member, *Brachyury,* is prominently expressed in midline mesoderm, especially notochord) contains five subfamilies, with *Tbr1* exhibiting a predominant forebrain expression (Bulfone et al. 1995; Papaioannou 2001). Like *reelin*, *Tbr1* is exclusively expressed in postmitotic cortical neurons, that is in Cajal-Retzius cells and early born (only glutamatergic, not GABAergic) neurons of layer 6 (later more weakly in all layers; Dwyer and O'Leary 2001; Hevner et al. 2001). In the *Tbr1* mutant mouse, *reelin* expression is greatly reduced in Cajal-Retzius cells and massive defects in early cortical migration/differentiation (layer 6) and connectivity emerge (Hevner et al. 2001). Similar to *Tbr1*, the *Emx2* gene—one of two vertebrate Emx-family genes representing homologs of *empty spiracles* in *Drosophila* (Simeone et al. 1992b)—is also expressed in Cajal-Retzius cells. The paralog *Emx1* is expressed in all other cortical layers during development. Both *Emx* genes are furthermore expressed in the proliferative ventricular zone of the cortex (Cecchi and Boncinelli 2000). Consistent with a role of *Emx2* in cortex development, *Emx2* null mutant mice show reduced numbers of Cajal-Retzius cells and an abnormal cortical migration pattern (Mallamaci et al. 2000). In contrast, the paired-box family gene *Pax6* is predominantly expressed in mitotic ventricular zone cells of the cortex (Walther and Gruss 1991; Stoykova and Gruss 1994; Puelles et al. 2000). There, *Pax6* controls the development of asymmetrically dividing radial glia cells that provide glutamatergic neurons of all cortical layers (Götz et al. 1998, 2002; Heins et al. 2002). More specifically, the *Pax6* gene is involved with regional specification of progenitor cells in the ventrolateral cortex (Stoykova et al. 2000; Toresson et al. 2000; Yun et al. 2001).

In addition, there is a migratory stream of *Pax6* positive, postmitotic, non-radial glia cells in the mouse telencephalon that extends peripherally towards the pia in the region of the pallial–subpallial boundary. This migratory stream has been interpreted alternatively as lying within the recently newly defined ventral pallium (which gives rise to the mammalian pallial amygdala, claustrum, and endopiriform area or to part of the sauropsidian DVR; Smith-Fernandez et al. 1998) or as the most dorsal aspect of the striatal formation (Puelles et al. 1999, 2000). Dorsal to this *Pax6* cell stream, some pallial genes, like *Emx1*, cease to be expressed, leaving a gap towards the striatal *Dlx* gene expressions (Smith-Fernandez et al. 1998). As all pallial divisions, the ventral pallium is further selectively characterized by *Tbr1*-expression (fig. 16B; Puelles et al. 1999, 2000), as is also part of the subpallial septum (fig. 16B).

The mouse subpallium exhibits distinct expressions of two paralogous homeobox genes of the Dlx family, i.e., *Dlx1* and *2* (homologs of *distal-less* in *Drosophila*) in the lateral (future striatum) and medial (future pallidum) ganglionic eminences, as well as in the septum (while these genes are not expressed in the cortex; fig. 16B; Bulfone et al. 1993a,b 1995; Eisenstat et al. 1999; Puelles et al. 2000). There is a strong implication of a function of *Dlx* genes with the development of GABAergic neurons, both of striatal projection neurons and of interneurons migrating dorsally into cortex (Anderson et al. 1997a,b; Eisenstat et al. 1999; Schuurmans and Guillemot 2002; Yun et al. 2002). The subpallial expression of *Dlx1/2* genes overlaps essentially with that of *Mash1* (see above) and, not surprisingly, *Mash1* (early) and *Dlx1/2* (later) act sequentially in the differentiation of subpallial GABAergic cells (Yun et al. 2002). The homeobox gene *Nkx2.1* (Casarosa et al. 1999; Sussel et al. 1999; Marín et al. 2000; Puelles et al. 2000) has a more restricted subpallial expression in the medial—but not lateral—ganglionic eminence and part of the septum. While the lateral ganglionic eminence is the site of origin of the local GABAergic striatal projection neurons, the medial ganglionic eminence produces GABAergic neurons migrating into striatum and cortex, and cholinergic neurons migrating into striatum and basal forebrain (Marín et al. 2000; Zhao et al. 2003). The *Nkx2.1* gene is critical for the expression of the LIM-homeobox genes *Lhx6* and *Lhx7/8* in this pathway (Sussel et al. 1999). There is evidence that *Lhx7* (= Lhx8; Zhao et al. 2003) and *Lhx6* differentially regulate the development of cholinergic or GABAergic neurons, respectively (Márin et al. 2000). *Lhx7/8* may have an additional role in GABAergic neuronal development (Zhao et al. 2003).

How is the situation in the zebrafish telencephalon in comparison? Although functional information is largely missing, comparable gene expression patterns in the zebrafish telencephalon are amazingly similar. We have already discussed the complementarity of pallial *neurog1/neurod* versus subpallial *Zash1a* expression (fig. 17A). Moreover, zebrafish *Dlx1/2* orthologs also show expression domains restricted to the subpallial (presumptive striatal and septal) ventral telencephalic area (fig. 17A; Zerucha et al. 2000). Of two *Nkx2.1* zebrafish paralogs, *Nkx2.1b* clearly is expressed in ventral—and not dorsal—telencephalon, but more detailed allocations within area ventralis cannot be drawn from the wholemount data presented (e.g., Rohr et al. 2001). The *sonic hedgehog* gene (*Shh*) is longitudinally expressed in the basal aspect of the brain and its activity is required for the induction of *Nkx2.1b* and also for *Nkx2.2*, but not for *Nkx2.1a* (Barth and Wilson 1995; Rohr et al. 2001). Therefore, subpallial gene expression patterns altogether indicate that the same genetic pathway related to the generation of GABAergic and maybe cholinergic neurons likely act both in teleostean and mammalian subpallium.

As in the mouse pallium, there are orthologous zebrafish genes in addition to the discussed bHLH genes (*neurogenin1*, *neurod*, fig. 17A) which characterize the dorsal (pallial)—but not ventral (subpallial)—telencephalic territory, i.e., *reelin* (Costagli et al. 2002), *emx1/2* (fig. 17A; Morita et al. 1995) and *Tbr1* (fig. 17A; Mione et al. 2001; note the additional subpallial, i.e., septal, expression and that in the eminentia thalami, see below). Furthermore, a detailed study of Pax6 protein expression (Wullimann and Rink 2001; fig. 17A) has revealed that the major telencephalic *Pax6* expression domain in the zebrafish corresponds to the migrating stream noted at the pallial–subpallial boundary in amniotes (see above). In contrast, a ventricularly located *Pax6* expression in proliferative radial glia cells as in the mouse pallium (i.e., isocortex; Götz et al. 1998) is not seen in the developing teleostean pallium (fig. 17A). Therefore, some developmental functions of *Pax6* in the zebrafish forebrain—such as inhibition of subpallial cell migration into the pallium by the *Pax6* migrating stream (Chapouton et al. 1999) or regional specification of progenitor cells in the ventrolateral cortex (Stoykova et al. 2000; Toresson et al. 2000; Yun et al. 2001)—might be shared with amniotes and maybe all vertebrates. But other *Pax6* functions—such as control of intrapallial radial migration of neurons via *Pax6* expressing radial glia cells—may be characteristic of amniotes or mammals only. Such fine, but distinct, differences in zebrafish gene expressions may be important clues to the understanding of how regulatory and other factors act in the zebrafish pallium where no cortex develops as in the mouse. Also zebrafish *reelin* expression in the pallium differs in comparison to the mammalian cortex. Although zebrafish *reelin* also is exclusively expressed in postmitotic cells, it is apparently not restricted to the earliest born cortical (Cajal-Retzius) cells as in the mouse. From a comparison of early and late zebrafish stages shown by Costagli et al. (2002), it appears that zebrafish *reelin* expression is continually present in newly born pallial cells immediately lateral to the proliferative ventricular zone over a long developmental period, but that *reelin* transcripts disappear in more peripheral differentiating pallial neurons. Although *reelin*-positive cells comparable to the mammalian Cajal-Retzius population have been described in the marginal zone of the sauropsid cortex (Tissir et al. 2003), *reelin* expression also occurs to a varying degree in deeper postmitotic cortical cell populations in certain sauropsids. Therefore, the function of *reelin* may be related to different morphogenetic developmental patterns in the pallium of vertebrates.

The migrating *Pax6* cell stream in the zebrafish telencephalon definitely lies at the subpallial/pallial boundary zone as in the mouse, but in order to resolve its exact location with respect to that of the ventral pallium and striatum, more refined gene expression data (e.g., exact delimitation of *Emx1* and *Tbr1* expressions) are needed in the zebrafish. More posteriorly in the early zebrafish telencephalon, Pax6 cells definitely extend increasingly more into the pallium, and this is the area where the developing (non-everted) lateral pallium is located and where the early pallial amygdala and other ventral pallial derivatives might be suspected (see Wullimann and Mueller 2004b). Interestingly, the latter territory (the most mediobasal part of Dm) has been identified functionally in the adult goldfish after lesion studies as amygdala (Portavella et al. 2002, 2003; Salas et al. 2003). Thus, similar to amniotes, the *Pax6* migratory stream would lie in the general region of the newly recognized ventral pallium (e.g., pallial amgydala) and lateral pallium (see Wullimann and Mueller 2004b for more extensive discussion). Similarly, future finer analysis in the zebrafish will have to show whether *Pax6* expressing cells extend into the striatal territory as has been reported for amniotes (mouse/chicken; Puelles et al.; 2000).

4.3. Comparable Early Diencephalic Gene Expression in Vertebrates

The correlated expression of the bHLH gene *Mash1* (Lo et al. 1991; Ma et al. 1997; Horton et al. 1999; Torii et al. 1999; Fode et al. 2000) and of *Dlx1/2* (Bulfone et al. 1993a,b) in the subpallium is furthermore seen in the mouse diencephalon in the ventral thalamus, and parts of preoptic region (fig. 16B) and hypothalamus (not shown). Also in the zebrafish, the same correlation of expression of the orthologous genes *Zash1a* (Wullimann and Mueller 2002) and *Dlx1/2* (Hauptmann and Gerster 2000; Zerucha et al. 2000) is present in ventral thalamus, preoptic region, and hypothalamus (fig. 17). As in the subpallium of both mouse and zebrafish, these genes may have roles in the specification of particular neurotransmitter phenotypes (e.g., likely GABAergic neurons in the ventral thalamus). Certain cellular co-localizations of *Dlx1/2* and *Mash1* (*Zash1a* in zebrafish) expressing cells with additional genetic markers are of interest in this context, i.e., with *Pax6* only in the ventral thalamus and with *Nkx2.1* (*Nkx2.1a* in zebrafish; Rohr et al. 2001) only in the hypothalamus. A role of these genes in the specification of dopaminergic (ventral thalamic) zona incerta cells has been indicated (Andrews et al. 2003), because DLX proteins co-occur with *Mash1* and *Pax6* in dopaminergic progenitor cells of the mouse alar plate ventral thalamus and *Dlx1/2* double mouse mutants do neither express *Pax 6* nor tyrosine hydroxylase in the ventral thalamus. This link of the *Pax6* gene to the development of ventral thalamic dopaminergic cells has been proposed based on gene expression patterns and dopamine cell distribution alone (Wullimann and Mueller 2002).

Expression patterns of *Pax6* in the alar plate diencephalon are highly similar in mouse (Walther and Gruss 1991; Stoykova and Gruss 1994) and zebrafish (Wullimann and Rink 2001; see discussion there for zebrafish particularities of *Pax6* expression relating to migrating diencephalic M1/M2 areas). All three alar plate prosomeric divisions (pretectum, dorsal thalamus, and ventral thalamus) show *Pax6* positive cells along the ventricle, but these domains differ in their lateral extent (e.g., massive laterally extending ventral thalamic domain; figs. 16 and 17), highlighting their respective prosomeric identities very similarly in mouse and zebrafish.

The mouse *Emx1/2* genes both have a cortical expression (see above; fig. 16B), with *Emx2* additionally exhibiting epithalamic, basal dorsal thalamic and hypothalamic domains (not shown; Simeone et al. 1992b). The initial description of two zebrafish *emx* genes delivered a highly similar picture: *emx1* has exclusively a dorsal telencephalic (pallial) expression (fig. 17A), *emx2* in addition a hypothalamic and a basal dorsal thalamic one (not shown; Morita et al. 1995), directly posterior to the zona limitans intrathalamica (Mathieu et al. 2002). Recently, Kawahara and Dawid (2002) described a different *emx1* gene with pallial, epithalamic, and basal dorsal thalamic (but not hypothalamic) expression domains and suggested that the initially described zebrafish *emx1* gene should be renamed *emx3*. Interestingly, three *emx* genes with similar expression domains are also present in the dogfish *Scyliorhinus canicula* (Derobert et al. 2002), representing a member of an outgroup of mammals and teleosts.

Regarding the diencephalon, many molecular markers discussed above could recently be used fruitfully in the zebrafish to increase the resolution of early molecular neuroanatomy (Wullimann and Mueller 2004a). The genes *Mash1* (Ma et al. 1997; Horton et al. 1999; Torii et al. 1999; Fode et al. 2000), *Pax6* (Walther and Gruss 1991; Stoykova and Gruss 1994; Puelles et al. 2000), *Dlx2* (Bulfone et al. 1993a,b, 1995; Eisenstat et al. 1999; Puelles et al. 2000) and *Tbr1* (Bulfone et al. 1995; Puelles et al. 2000) as well as *Lhx9* (Rétaux et al. 1999) are very selectively expressed in the fetal mouse ventral thalamus (*Dlx2*, *Mash1*, *Pax6*), eminentia thalami (*Lhx9*, *Tbr1*, *Pax6*-restricted to ventricular zone) or preoptic divisions (*Dlx2*, *Mash1*, *Pax6*-restricted to caudolateral part). A comparative approach allowed us to newly identify in the zebrafish the early eminentia thalami and to characterize it as a *neurog1*/*neurod* positive area, which likely gives rise to the *Tbr1*-positive (Mione et al. 2001) neurons destined to become the entopeduncular/ peripeduncular complex (Wullimann and Mueller 2004a).

4.4. Chick and Frog Forebrain Patterns

Expression domains of the homologs of the genes under consideration in the other major developmental amniote model, the chicken, (i.e., *Cash1, ngn1, Emx1/2, Pax6, Dlx1/2,* Tbr1; Jasoni et al. 1994; Smith-Fernandez et al. 1998; Puelles et al. 2000; von Frowein et al. 2002) generally reveal a picture conforming very well to that in the mouse. The relevance for a comparative understanding of avian and mammalian forebrain gene expression patterns has been extensively documented and discussed elsewhere (Smith-Fernandez et al. 1998; Puelles et al. 2000; von Frowein et al. 2002).

Furthermore, the best established amphibian model animal, the clawed frog *Xenopus laevis*, has been crucial in the elucidation of the roles of the genes involved in neurogenesis under consideration in primary neurogenesis. We have discussed this information in Chapter 2 at some length, as it forms an important foundation for the present book. The available information regarding pallial versus subpallial gene expressions in *Xenopus* is highly consistent with the situation in amniotes, showing, for example, clear pallial expression of *Emx1* and complementary subpallial expression of *Dlx1* (*xDll3*) and *Dlx2* (*xDll4*) (Pannese et al. 1998; Smith-Fernandez et al. 1998; Bachy et al. 2001, 2002; Brox et al. 2004). However, the later expression and roles of most genes under consideration still awaits proper documentation in the entire developing frog brain, especially regarding *neurogenin* and *neuroD*.

In conclusion, the analysis presented in this chapter shows that our previously unattained level of resolution in the definition of many finer subdivisions in the zebrafish forebrain may be put fruitfully into context with information in other vertebrate models, especially the well investigated mouse model. The establishment of such relationships will help for using the zebrafish in more neurobiologically oriented genetic studies as well as in the improvement of the understanding of the developmental basis of vertebrate brain evolution.

List of Abbreviations

ac	anterior commissure
ALLG	anterior lateral line ganglion
AP	alar plate
b	basal tectal proliferation
BP	basal plate
CC	cerebellar crest
cec	cerebellar commissure
CeP	cerebellar plate
Ch	chorda dorsalis
DIL	diffuse nucleus of inferior lobe
DP	dorsal pallium (isocortex)
DT	dorsal thalamus (thalamus)
DVe	diencephalic ventricle
E	epiphysis
EG	eminentia granularis
EGL	external granular layer
EmT	eminentia thalami
EPi	eye pigment
FG	facial ganglion
FP	floor plate
fr	fasciculus retroflexus
GG	glossopharyngeal ganglion
Gl	olfactory bulb glomeruli
GT	griseum tectale
H	hypothalamus
Ha	habenula
hac	habenular commissure
Hc	caudal hypothalamus
Hi	intermediate hypothalamus
Hr	rostral hypothalamus
Hy	hypophysis (pituitary)
IMR	intermediate raphe
IMRF	intermediate reticular formation
IO	inferior olive
IR	inferior raphe
IRF	inferior reticular formation
l	lateral tectal proliferation
lfb	lateral forebrain bundle
LGE	lateral ganglionic eminence
LHP	lateral hinge-point between cerebellum/medulla oblongata
LP	lateral (olfactory) pallium
LV	lateral ventricle (mouse)
LVe	lateral recess ventricle of H
M1	migrated pretectal area
M2	migrated posterior tubercular area
M3	migrated area of EmT
M4	telencephalic migrated area
m	medial tectal proliferation
MCP	medial cerebellar proliferation
md	dorsal part of m
Mes	mesencephalon
MGE	medial ganglionic eminence
MHB	midbrain-hindbrain boundary
mlf	medial longitudinal fascicle
MN	Mauthner neuron
MO	medulla oblongata
MP	medial pallium (hippocampus)
mv	ventral part of m
NIII	oculomotor nerve nucleus
NIIIn	oculomotor nerve
N	region of the nucleus of medial longitudinal fascicle
NIn	nucleus interpeduncularis
OB	olfactory bulb
oc	optic chiasma
OC	otic capsule
OE	olfactory epithelium
OG	octaval ganglion
on	optic nerve
OR	optic recess
P1-P6	prosomeres 1-6
P	pallium
pc	posterior commissure
Pi	pigment
PLLG	posterior lateral line ganglion
Po	preoptic region
poc	postoptic commissure
Pr	pretectum
PT	posterior tuberculum
PTd	dorsal part of posterior tuberculum
PTM	posterior tectal membrane
PTv	ventral part of posterior tuberculum
PVe	posterior recess ventricle of H
RCT	rostral cerebellar thickening (valvula)
RF	rhombic fossa
R1-R7	rhombomeres 1-7
Rho-AP	rhombencephalic alar plate proliferation
Rho-BP	rhombencephalic basal plate proliferation
RL	rhombic lip
RP	roof plate
RVe	rhombencephalic ventricle
S	subpallium
SCO	subcommissural organ
Sd	dorsal division of S
Sep	septum
sm	somatomotor column
SP	secondary prosencephalon
SR	superior raphe
SRF	superior reticular formation
ss	somaotosensory column
Sv	ventral division of S
T	midbrain tegmentum (telencephalon in Figs. 1; 16)
TeO	tectum opticum
TeVe	tectal ventricle
TG	trigeminal ganglion
TL	torus longitudinalis
TLa	torus lateralis
TS	torus semicircularis
TVe	telencephalic ventricle
3V	diencephalic ventricle (Figure 16)
Va	valvula cerebelli
VCP	ventral cerebellar proliferative layer
Ve	brain ventricle
VG	vagal ganglion
vm	visceromotor column
VP	ventral pallium
vs	viscerosensory column
VT	ventral thalamus (prethalamus)
ZLI	zona limitans intrathalamica

References

Allende ML, Weinberg ES (1994) The expression pattern of two zebrafish *achaete-scute* homolog (*ash*) genes is altered in the embryonic brain of the *cyclops* mutant. *Dev. Biol., 166*, 509-530.

Altman J, Bayer SA (1979) Development of the diencephalon in the rat. VI. Re-evaluation of the embryonic development of the thalamus on the basis of thymidine-radiographic datings. *J. Comp. Neurol., 188*, 501-524.

Altman J, Bayer SA (1988) Development of the rat thalamus: I. Mosaic organization of the thalamic neuroepithelium. *J. Comp. Neurol., 275*, 346-377.

Andermann P, Ungos J, Raible DW (2002) *Neurogenin1* defines zebrafish sensory ganglia precursors. *Dev. Biol., 251*, 45-58.

Anderson SA, Eisenstat DD, Rubenstein JL (1997a) Interneuron migration from basal forebrain to neocortex: dependence on *Dlx* genes. *Science, 278*, 474-476.

Anderson SA, Qiu M, Bulfone A, Eisenstat DD, Meneses J, Pedersen R, Rubenstein JL (1997b) Mutations of the homeobox genes *Dlx-1* and *Dlx-2* disrupt the striatal subventricular zone and differentiation of the late born striatal neurons. *Neuron, 19*, 27-37.

Andrews GL, Yun K, Rubenstein JLR, Mastick GS (2003) Dlx transcription factors regulate development of dopaminergic neurons of the ventral thalamus. *Mol. Cell. Neurosci., 23*, 107-120.

Angevine JB Jr. (1970) Time and neuron origin in the diencephalon of the mouse. An autoradiographic study. *J. Comp. Neurol., 139*, 129-188.

Angevine JB Jr., Sidman RL (1961) Autoradiographic study of cell migration during histogenesis of cerebral cortex in the mouse. *Nature, 192*, 766-768.

Appel B, Eisen JS (1998) Regulation of neuronal specification in the zebrafish spinal cord by Delta function. *Development, 125*, 371-380.

Appel B, Givan LA, Eisen LA (2001) Delta-Notch signalling and lateral inhibition in zebrafish spinal cord development. *BMC Dev. Biol., 1*, 13.

Bachy I, Vernier P, Rétaux S (2001) The LIM-homeodomain gene family in the developing *Xenopus* brain: conservation and divergences with the mouse related to the evolution of the forebrain. *J. Neurosci., 21*, 7620-7629.

Bachy I, Berthon J, Rétaux S (2002) Defining pallial and subpallial divisions in the developing *Xenopus* forebrain. *Mech. Dev., 117*, 163-172.

Bally-Cuif L, Wassef M (1995) Determination events in the nervous system of the vertebrate embryo. *Curr. Opin. Genet. Dev., 5*, 450-458.

Bally-Cuif L, Dubois L, Vincent A (1998) Molecular cloning of *Zcoe2*, the zebrafish homolog of Xenopus *Xcoe2* and mouse *EBF-2*, and its expression during primary neurogenesis. *Mech. Dev., 77*, 85-90.

Bar I, Lambert de Rouvroit C, Goffinet AM (2000) The evolution of cortical development. An hypothesis based on the role of the Reelin signaling pathway. *TINS, 23*, 633-638.

Barami K, Iversen K, Furneaux H, Goldman SA (1995) Hu protein as an early marker of neuronal phenotypic differentiation by subependymal zone cells of the adult songbird forebrain. *J. Neurobiol., 28*, 82-101.

Barth KA, Wilson SW (1995) Expression of zebrafish *nk2.2* is influenced by *sonic hedgehog/vertebrate hedgehog-1* and demarcates a zone of neuronal differentiation in the embryonic forebrain. *Development, 121*, 1755-1768.

Bayer SA, Altman J (1974) Hippocampal development in the rat: cytogenesis and morphogenesis examined with autoradiography and low-level X-irradiation. *J. Comp. Neurol., 158*, 55-79.

Bayer SA, Altman J (1987) Directions in neurogenetic gradients and patterns of anatomical connections in the telencephalon. *Prog. Neurobiol., 29*, 57-106.

Bayer SA, Altman J (1995a) Neurogenesis and neuronal migration. In: Paxinos GT (Ed.), *Rat Nervous System*, 2nd Ed. pp. 1041-1078. Academic Press, San Diego.

Bayer SA, Altman J (1995b) Principles of neurogenesis, neuronal migration and neural circuit formation. In: Paxinos GT (Ed.), *Rat Nervous System*, 2nd Ed. pp. 1079-1098. Academic Press, San Diego.

Bergquist H (1932) Zur Morphologie des Zwischenhirns bei niederen Wirbeltieren. *Acta Zool. (Stockholm), 13*, 57-304.

Bergquist H (1954) *Ontogenesis of Diencephalic Nuclei in Vertebrates.* Lunds Universitets Arsskrift. N.F. Avd. 2. Bd 50. Nr. 6 (Kungl. Fysiografiska Sällskapets Handlingar. N.F. Bd 65, Nr. 6): 1-34.

Bergquist H, Källén B (1954) Notes on the early histogenesis and morphogenesis of the central nervous system in vertebrates. *J. Comp. Neurol., 100*, 627-659.

Bernhardt RR, Chitnis AB, Lindamer L, Kuwada JY (1990) Identification of spinal neurons in the embryonic and larval zebrafish. *J. Comp. Neurol., 302*, 603-616.

Bierkamp C, Campos-Ortega JA (1993) A zebrafish homologue of the *Drosophila* neurogenic gene *Notch* and its pattern of transcription during early embryogenesis. *Mech. Dev., 43*, 87-100.

Birgbauer E, Fraser SE (1994) Violation of cell lineage restriction compartments in the chick hindbrain. *Development, 120*, 1347-1356.

Blader P, Fischer N, Gradwohl G, Guillemot F, Strähle U (1997) The activity of *neurogenin1* is controlled by local cues in the zebrafish. *Development, 124*, 4557-4569.

Boncinelli E, Gulisano M, Broccoli V (1993) *Emx* and *Otx* homeobox genes in the developing mouse brain. *J. Neurobiol., 24*, 1356-1366.

Boncinelli E, Gulisano M, Spada F, Broccoli V (1995) Emx and Otx Gene expression in the developing mouse brain. In: Bock GR, Cardew G (Eds.), *Development of the Cerebral Cortex, Ciba Foundation Symposium 193.* Wiley, Chichester.

Brand M, Heisenberg CP, Jiang YJ, Beuchle D, Lun K, Furutani-Seiki M, Granato M, Haffter P, Hammerschmidt M, Kane DA, Kelsh RN,

Mullins MC, Odenthal J, van Eeden FJ, Nüsslein-Volhard C (1996) Mutations in zebrafish genes affecting the formation of the boundary between midbrain and hindbrain. *Development, 123*, 179-190.

Brox A, Puelles L, Ferreiro B, Medina L (2004) Expression of the genes *Emx1, Tbr1, Eomes* (*Tbr2*) in the telencephalon of *Xenopus laevis*, confirms the existence of a ventral pallial division in all tetrapods. *J. Comp. Neurol., 474*, 562-577.

Bulfone A, Kim H-J, Puelles L, Porteus MH, Grippo JF, Rubenstein JLR (1993a) The mouse *Dlx-2* (*Tes-1*) gene is expressed in spatially restricted domains of the forebrain, face and limbs in midgestation mouse embryos. *Mech. Dev., 40*, 129-140.

Bulfone A, Puelles L, Porteus MH, Frohman MA, Martin GR, Rubenstein JLR (1993b) Spatially restricted expression of *Dlx-1*, *Dlx-2* (*Tes-1*), *Gbx-2*, and *Wnt-3* in the embryonic day 12.5 mouse forebrain defines potential transverse and longitudinal segmental boundaries. *J. Neurosci., 13*, 3155-3172.

Bulfone A, Smiga SM, Shimamura K, Peterson A, Puelles P, Rubenstein JLR (1995) *T-Brain-1*: a homolog of *Brachyury* whose expression defines molecularly distinct domains within the cerebral cortex. *Neuron, 15*, 63-78.

Campbell K (2003) Dorsal–ventral patterning in the mammalian telencephalon. *Curr. Opin. Neurobiol., 13*, 50-56.

Casarosa S, Fode C, Guillemot F (1999) *Mash1* regulates neurogenesis in the ventral telencephalon. *Development, 126*, 525-534.

Cecchi C, Boncinelli E (2000) *Emx* homeogenes and mouse brain development. *TINS, 23*, 347-352.

Chapouton P, Gärtner A, Götz M (1999) The role of *Pax6* in restricting cell migration between developing cortex and basal ganglia. *Development, 126*, 5569-5579.

Chitnis AB (1999) Control of neurogenesis—lessons from frogs, fish and flies. *Curr. Opin. Neurobiol., 9*, 18-25.

Chitnis A, Kintner C (1996) Sensitivity of proneural genes to lateral inhibition affects the pattern of primary neurogenesis in Xenopus embryos. *Development, 122*, 2295-2301.

Chitnis AB, Kuwada JY (1990) Axonogenesis in the brain of zebrafish embryos. *J. Neurosci., 10*, 1892-1905.

Chitnis A, Henrique D, Lewis J, Ish-Horowicz D, Kintner C (1995) Primary neurogenesis in *Xenopus* embryos regulated by a homologue of the *Drosophila* neurogeneic gene *Delta*. *Nature, 375*, 761-766.

Clarke JDW, Lumsden A (1993) Segmental repetition of neuronal phenotypes in the chick embryo. *Development, 118*, 151-162.

Coffman C, Harris W, Kintner C (1990) Xotch, the *Xenopus* homolog of *Drosophila* notch. *Science, 249*, 1438-1441.

Coffman CR, Skoglund P, Harris W, Kintner C (1993) Expression of an extracellular deletion of Xotch diverts cell fate in *Xenopus* embryos. *Cell, 73*, 659-671.

Costagli A, Kapsimali M, Wilson SW, Mione M (2002) Conserved and divergent patterns of *reelin* expression in the zebrafish central nervous system. *J. Comp. Neurol., 450*, 73-93.

Curran T, D'Arcangelo G (1998) Role of *reelin* in the control of brain development. *Brain Res. Rev., 26*, 285-294.

Derobert Y, Plouhine JL, Sauka-Spengler T, Le Mentec C, Baratte B, Jaillard D, Mazan S (2002) Structure and expression of three *Emx* genes in the dogfish *Scyliorhinus canicula*: functional and evolutionary implications. *Dev. Biol., 247*, 390-404.

Dornseifer P, Takke C, Campos-Ortega JA (1997) Overexpression of a zebrafish homologue of the *Drosophila* neurogenic gene *Delta* perturbs differentiation of primary neurons and somite development. *Mech. Dev., 63*, 159-171.

Dorsky RI, Rapaport DH, Harris WA (1995) Xotch inhibits cell differentiation in the *Xenopus* retina. *Neuron, 14*, 487-496.

Dorsky RI, Chang WS, Rapaport DH, Harris WA (1997) Regulation of neuronal diversity in the *Xenopus* retina by Delta signalling. *Nature, 385*, 67-70.

Dubbeldam JL (1998) Birds. In: Nieuwenhuys R, ten Donkelaar HJ, Nicholson C (Eds.), *The Central Nervous System of Vertebrates*, Vol. 3, pp. 1525-1636. Springer, New York.

Dwyer ND, O'Leary DDM (2001) *Tbr1* conducts orchestration of early cortical development. *Neuron, 29*, 309-311.

Edwards MA, Yamamoto M, Cavniess VS Jr. (1990) Organization of radial glia and related cells in the developing murine CNS. An analysis based upon a new monoclonal antibody marker. *Neuroscience, 36*, 121-144.

Eisenstat DD, Liu JK, Mione M, Zhong W, Yu G, Anderson SA, Ghattas I, Puelles P, Rubenstein JLR (1999) DLX-1, DLX-2, and DLX-5 expression define distinct stages of basal forebrain differentiation. *J. Comp. Neurol., 414*, 217-237.

Ericson J, Muhr J, Placzek M, Lints T, Jessell TM, Edlund T (1995) Sonic hedgehog induces the differentiation of ventral forebrain neurons: a common signal for ventral patterning within neural tube. *Cell, 81*, 747-756.

Figdor MC, Stern CD (1993) Segmental organization of embryonic diencephalon. *Nature, 363*, 630-634.

Fode C, Ma Q, Casarosa S, Ang S-L, Anderson DJ, Guillemot F (2000) A role for neural determination genes in specifying the dorsoventral identity of telencephalic neurons. *Genes Dev., 14*, 67-80.

Fraser S, Keynes R, Lumsden A (1990) Segmentation in the chick embryo hindbrain is defined by cell lineage restrictions. *Nature, 344*, 431-435.

Fujita S (1963) The matrix cell and cytogenesis in the developing central nervous system. *J. Comp. Neurol., 120*, 37-42.

Fujita S (1964) Analysis of neuron differentiation in the central nervous system by tritiated thymidine autoradiography. *J. Comp. Neurol., 122*, 311-327.

Fujita S (1966) Application of light and electron microscopic autoradiography to the study of cytogenesis of the forebrain. In: Hassler R, Stephan H (Eds.), *Evolution of the Forebrain*, pp. 180-196. Thieme, Stuttgart.

Garcia-Calero E, Martinez-de-la-Torre M, Puelles L (2002) The avian griseum tectale: cytoarchitecture, NOS expression and neurogenesis. *Brain Res. Bull., 57*, 353-357.

Gaskell WH (1889) On the relation between the structure, function, distribution and origin of the cranial nerves; together with a theory of the origin of the nervous system of vertebrata. *J. Physiol., 10*, 153-211.

Götz M, Stoykova A, Gruss P (1998) *Pax6* controls radial glia differentiation in the cerebral cortex. *Neuron, 21*, 1031-1044.

Götz M, Hartfuss E, Malatesta P (2002) Radial glia cells as neuronal precursors: a new perspective on the correlation of morphology and lineage restriction in the developing cerebral cortex of mice. *Brain Res. Bull.*, *57*, 777-788.

Goldowitz D, Hamre K (1998) The cells and molecules that make up a cerebellum. *TINS*, *21*, 375-382.

Gong Z, Ju B, Wan H (2001) Green fluorescent protein (GFP) transgenic fish and their application. *Genetica*, *111*, 213-225.

Graham A, Papalopulu N, Krumlauf R (1989) The murine and *Drosophila* homeobox gene complexes have common features of organization and expression. *Cell*, *57*, 367-378.

Guillemot G, Joyner AL (1993) Dynamic expression of the murine *Achaete-Scute* homologue *Mash-1* in the developing nervous system. *Mech. Dev.*, *42*, 171-185.

Hack I, Bancila M, Loulier K, Carroll P, Cremer H (2002) Reelin is a detachment signal in tangential chain-migration during postnatal neurogenesis. *Nature Neurosci.*, *5*, 939-945.

Haddon C, Smithers L, Schneider-Maunoury S, Coche T, Henrique D, Lewis L (1998) Multiple *delta* genes and lateral inhibition in zebrafish primary neurogenesis. *Development*, *125*, 359-370.

Hannemann E, Trevarrow B, Metcalfe WK, Kimmel CB, Westerfield M (1988) Segmental pattern of development in the hindbrain and spinal cord of the zebrafish embryo. *Development*, *103*, 49-58.

Hartfuss E, Galli R, Heins N, Götz M (2001) Characterization of CNS precursor subtypes and radial glia. *Dev. Biol.*, *229*, 15-30.

Hatten ME (1990) Riding the glial monorail: a common mechanism for glial-guided neuronal migration in different regions of the developing mammalian brain. *TINS*, *13*, 179-184.

Hatten ME (1993) The role of migration in CNS neuronal development. *Curr. Opin. Neurobiol.*, *3*, 38-44.

Hatten ME (1999) Central nervous system neuronal migration. *Annu. Rev. Neurosci.*, *22*, 511-539.

Hatten ME (2002) New directions in neuronal migration. *Science*, *297*, 1660-1663.

Hatten ME, Alder J, Zimmerman K, Heintz N (1997) Genes involved in cerebellar cell specification and differentiation. *Curr. Opin. Neurobiol.*, *7*, 40-47.

Hauptmann G, Gerster T (2000) Regulatory gene expression patterns reveal transverse and longitudinal subdivisions of the embryonic zebrafish forebrain. *Mech. Dev.*, *91*, 105-118.

He W, Ingraham C, Rising L, Goderie S, Temple S (2001) Multipotent stem cells from the mouse basal forebrain contribute GABAergic neurons and oligodendrocytes to the cerebral cortex during embryogenesis. *J. Neurosci.*, *21*, 8854-8862.

Heins N, Malatesta P, Cecconi F, Nakafuku M, Tucker KL, Hack MA, Chapouton P, Barde Y-A, Götz M (2002) Glial cells generate neurons: the role of the transcription factor Pax6. *Nat. Neurosci.*, *5*, 308-315.

Herrick CJ (1910) The morphology of the forebrain in amphibia and reptilia. *J. Comp. Neurol.*, *20*, 413-547.

Hevner RF, Shi L, Justice N, Hsueh Y-P, Sheng M, Smiga S, Bulfone A, Goffinet AM, Campagnoni AT, Rubenstein JLR (2001) *Tbr1* regulates differentiation of the preplate and layer 6. *Neuron*, *29*, 353-366.

Hirsch MR, Tiveron MC, Guillemot F, Brunet JF, Goridis C (1998) Control of noradrenergic differentiation and *Phox2a* expression by MASH1 in the central and peripheral nervous system. *Development*, *125*, 599-608.

His W (1888) Zur Geschichte des Gehirns sowie der centralen und periferischen Nervenbahnen beim menschlichen Embryo. *Abh. Math. Phys. Kl. Kgl. Sächs. Ges. Wiss.*, *14*, 339-393.

His W (1893a) Über das frontale Ende des Gehirnrohres. *Arch. Anat. Physiol. Anat Abt.*, 157-172.

His W (1893b) Vorschläge zur Eintheilung des Gehirns. *Arch. Anat. Physiol. Anat. Abt.*, 172-180.

Holland PWH, Hogan BLM (1988) Expression of homoebox genes during mouse development: a review. *Genes Dev.*, *2*, 773-782.

Horton S, Meredith A, Richardson JA, Johnson JE (1999) Correct coordination of neuronal differentiation events in ventral forebrain requires the bHLH factor Mash1. *Mol. Cell. Neurosci.*, *14*, 355-369.

Hunt P, Krumlauf R (1992) *Hox* codes and positional specification in vertebrate embyronic axes. *Annu. Rev. Cell Biol.*, *8*, 227-256.

Jasoni CL, Walker MB, Morrism MD, Reh TA (1994) A chicken acheate-scute homolog (CASH-1) is expressed in a temporally and spatially discrete manner in the developing nervous system. *Development*, *120*, 769-783.

Johnston JB (1909) The morphology of the forebrain vesicle in vertebrates. *J. Comp. Neurol.*, *19*, 457-539.

Källén B (1951) Embryological studies on the nuclei and their homologization in the vertebrate forebrain. *Kgl. Fysiogr. Sällsk. Lund Handl. N.F.*, *62* (5).

Källén B (1952) Notes on the proliferation processes in the neuromeres in vertebrate embryos. *Acta Soc. Med. Upsal.*, *57*, 111-118.

Kawahara A, Dawid IB (2002) Developmental expression of zebrafish *emx1* during early embryogenesis. *Gene Expr. Patterns*, *2*, 201-206.

Kawakami K, Amsterdam A, Shimoda N, Becker T, Mugg J, Shima A, Hopkins N (2000) Proviral insertions in the zebrafish *hagoromo* gene, encoding an F-box/WD40-repeat protein, cause stripe pattern anomalies. *Curr. Biol.*, *10*, 463-466.

Kerstetter AE, Azodi E, Marrs JA, Liu Q (2004) Cadherin-2 function in the cranial ganglia and lateral line system of developing zebrafish. *Dev. Dyn.*, *230*, 137-143.

Keynes RJ, Stern CD (1984) Segmentation in the vertebrate nervous system. *Nature*, *310*, 786-789.

Keynes RJ, Stern CD (1988) Mechanisms of vertebrate segmentation. *Development*, *103*, 413-429.

Kimmel CB (1993) Patterning the brain of the zebrafish embryo. *Annu. Rev. Neurosci.*, *16*, 707-732.

Kimmel CB, Warga RM, Kane DA (1994) Cell cycles and clonal strings during formation of the zebrafish central nervous system. *Development*, *120*, 265-276.

Kimmel CB, Ballard WW, Kimmel SR, Ullmann B, Schilling TF (1995) Stages of embryonic development of the zebrafish. *Dev. Dyn.*, *203*, 253-310.

Kingsbury BF (1922) The fundamental plan of the vertebrate brain. *J. Comp. Neurol.*, *34*, 461-491.

Komuro H, Rakic P (1998) Distinct modes of neuronal migration in different domains of developing cerebellar cortex. *J. Neurosci.*, *18*, 1478-1490.

Köster RW, Fraser SE (2001) Direct imaging of in vivo neuronal migration in the developing cerebellum. *Curr. Biol.*, *11*, 1858-1863.

Korzh V, Sleptsova I, Liao J, He J, Gong Z (1998) Expression of zebrafish bHLH genes *ngn1* and *nrd* defines distinct stages of neural differentiation. *Dev. Dyn.*, *213*, 92-104.

Larsen CW, Zeltser LM, Lumsden A (2001) Boundary formation and compartition in the avian diencephalon. *J. Neurosci.*, *21*, 4699-4711.

Lee J-K, Cho J-H, Hwang W-S, Lee J-D, Reu D-S, Suh-Kim H (2000) Expression of NeuroD/BETA2 in mitotic and postmitotic neuronal cells during the development of nervous system. *Dev. Dyn.*, *217*, 361-367.

Lewis L (1996) Neurogenic genes and vertebrate neurogenesis. *Curr. Opin. Neurobiol.*, *6*, 3-10.

Lewis J (1998) Multiple delta genes and lateral inhibition in zebrafish primary neurogenesis. *Development*, *125*, 359-370.

Li Z, Hu M, Ochocinska MJ, Joeph NM, Easter SS Jr. (2000) Modulation of cell proliferation in the embryonic retina of zebrafish (*Danio rerio*). *Dev. Dyn.*, *219*, 391-401.

Liao J, He J, Yan T, Korzh V, Gong Z (1999) A class of neuroD-related basic helix–loop–helix transcription factors expressed in developing central nervous system in zebrafish. *DNA Cell Biol.*, *18*, 333-344.

Liu Q, Ensign RD, Azodi E (2003) Cadherin-1, -2 and -4 expression in the cranial ganglia and lateral line system of developing zebrafish. *Gene Expr. Patterns*, *3*, 653-658.

Lo L-C, Johnson JE, Wuenschell CW, Saito T, Anderson DJ (1991) Mammalian *achaete-scute* homolog 1 is transiently expressed by spatially restricted subsets of early neuroepithelial and neural crest cells. *Genes Dev.*, *5*, 1524-1537.

Lo L, Tiveron MC, Anderson DJ (1998) MASH1 activates expression of the paired homeodomain transcription factor *Phox2a*, and couples pan-neuronal and subtype-specific components of autonomic neuronal identity. *Development*, *125*, 609-620.

Lumsden A (1990) The cellular basis of segmentation in the developing hindbrain. *TINS*, *13*, 329-339.

Lumsden A, Keynes R (1989) Segmental patterns of neuronal development in the chick hindbrain. *Nature*, *337*, 424-428.

Lumsden A, Krumlauf R (1996) Patterning the vertebrate neuraxis. *Science*, *274*, 1109-1115.

Luskin MB (1993) Restricted proliferation and migration of postnatally generated neurons derived from the forebrain subventricular zone. *Neuron*, *11*, 173-189.

Ma Q, Kintner C, Anderson DJ (1996) Identification of *neurogenin*, a vertebrate neuronal determination gene. *Cell*, *87*, 43-52.

Ma Q, Sommer L, Cserjesi P, Anderson DJ (1997) *Mash1* and *neurogenin1* expression patterns define complementary domains of neuroepithelium in the developing CNS and are correlated with regions expressing Notch ligands. *J. Neurosci.*, *17*, 3644-3652.

Malatesta P, Hartfuss E, Götz M (2000) Isolation of radial glial cells by fluorescent-activated cell sorting reveals a neuronal lineage. *Development*, *127*, 5253-5263.

Malatesta P, Hack MA, Hartfuss E, Kettenmann H, Klinkert W, Kirchhoff F, Götz M (2003) Neuronal or glial progeny: regional differences in radial glia fate. *Neuron*, *37*, 751-764.

Mallamaci A, Mercurio S, Muzio L, Cecchi C, Pardini CL, Gruss P, Boncinelli E (2000) The lack of *Emx2* causes impairment of *reelin* signaling and defects of neuronal migration in the developing cerebral cortex. *J. Neurosci.*, *20*, 1109-1118.

Marcus RC, Delaney CL, Easter SS Jr. (1999) Neurogenesis in the visual system of embryonic and adult zebrafish (*Danio rerio*). *Visual Neurosci.*, *16*, 417-424.

Marín F, Puelles L (1994) Patterning of the embryonic avian midbrain after experimental inversions: a polarizing activity from the isthmus. *Dev. Biol.*, *163*, 19-37.

Marín O, Anderson SA, Rubenstein JLR (2000) Origin and molecular specification of striatal interneurons. *J. Neurosci.*, *20*, 6063-6076.

Marusich MF, Furneaux HM, Henion PD, Weston JA (1994) Hu neuronal proteins are expressed in proliferating neurogenic cells. *J. Neurobiol.*, *25*, 143-155.

Mathews MB, Bernstein RM, Franza BR Jr., Garrels JI (1984) Identity of the proliferating cell nuclear antigen and cyclin. *Nature*, *309*, 374-376.

Mathieu J, Barth A, Rosa FM, Wilson SW, Peyriéras N (2002) Distinct and cooperative roles of Nodal and Hedgehog signal during hypothalamic development. *Development*, *129*, 3055-3066.

Mione M, Shanmugalingam S, Kimelman D, Griffin K (2001) Overlapping expression of zebrafish *T-Brain-1* and eomesodermin during forebrain development. *Mech. Dev.*, *100*, 93-97.

Morita T, Nitta H, Kiyama Y, Mori H, Mishina M (1995) Differential expression of two zebrafish *emx* homeoprotein mRNAs in the developing brain. *Neurosci. Lett.*, *198*, 131-134.

Mueller T, Wullimann MF (2002a) Expression domains of *neuroD* (*nrd*) in the early postembryonic zebrafish brain. *Brain Res. Bull.*, *57*, 377-379.

Mueller T, Wullimann MF (2002b) BrdU-, *neuroD-*(*nrd*) and Hu-studies show unusual non-ventricular neurogenesis in the postembryonic zebrafish forebrain. *Mech. Dev.*, *117*, 123-135.

Mueller T, Wullimann MF (2003) Anatomy of neurogenesis in the zebrafish brain. *Dev. Brain Res.*, *140*, 135-153.

Murphy P, Davidson DR, Hill RE (1989) Segment-specific expression of a homeobox-containing gene in the mouse hindbrain. *Nature*, *341*, 156-159.

Nieuwenhuys R (1998a) Morphogenesis and general structure. In: Nieuwenhuys R, ten Donkelaar HJ, Nicholson C (Eds.), *The Central Nervous System of Vertebrates*, Vol. 1, pp. 159-228. Springer, New York.

Nieuwenhuys R (1998b) Histogenesis. In: Nieuwenhuys R, ten Donkelaar HJ, Nicholson C (Eds.), *The Central Nervous System of Vertebrates*, Vol. 1, pp. 229-272. Springer, New York.

Nieuwenhuys R (1998c) Comparative neuroanatomy: place, principles and programme. In: Nieuwenhuys R, ten Donkelaar HJ, Nicholson C (Eds.), *The Central Nervous System of Vertebrates*, Vol. 1, pp. 273-326. Springer, New York.

Noctor SC, Flint AC, Weissman TA, Dammerman RS, Kriegstein AR (2001) Neurons derived from radial glial cells establish radial units in neocortex. *Nature*, *409*, 714-720.

O'Rourke N, Dailey ME, Smith SJ, McConell S (1992) Diverse migratory pathways in the developing cerebral cortex. *Science, 258*, 299-302.

Orr HA (1887) Contributions to the embryology of the lizard. *J. Morphol., 1*, 311-372.

Oxtoby E, Jowett T (1993) Cloning of the zebrafish *krox-20* gene (*krx-20*) and its expression during hindbrain development. *Nucleic Acids Res., 21*, 1087-1095.

Pannese M, Lupo G, Kabla B, Boncinelli E, Barsacchi G, Vignali R (1998) The *Xenopus Emx* genes identify presumptive dorsal telencephalon and are induced by head organizer signals. *Mech. Dev., 73*, 73-83.

Papaioannou VE (2001) T-Box genes in development: from hydra to humans. *Int. Rev. Cytol., 207*, 1-70.

Park HC, Hong SK, Kim HS, Kim SH, Yoon EJ, Kim CH, Miki N, Huh TL (2000) Structural comparison of zebrafish Elav/Hu and their differential expressions during neurogenesis. *Neurosci. Lett., 279*, 81-84.

Parnavelas JG (2000) The origin and migration of cortical neurones: new vistas. *TINS, 23*, 126-131.

Parras CM, Schuurmans C, Scardigli R, Kim J, Anderson DJ, Guillemot F (2002) Divergent functions of the proneural genes *Mash1* and *Ngn2* in the specification of neuronal subtype identity. *Genes Dev., 16*, 324-338.

Portavella M, Vargas JP, Torres B, Salas C (2002) The effects of telencephalic pallial lesions on spatial, temporal, and emotional learning in goldfish. *Brain Res. Bull., 57*, 397-399.

Portavella M, Salas C, Vargas JP, Papini MR (2003) Involvement of the telencephalon in spaced-trial avoidance learning in the goldfish (*Carassius auratus*). *Physiol. Behav., 80*, 49-56.

Price DJ, Aslam S, Tasker L, Gillies K (1997) Fates of the earliest generated cells in the developing murine cortex. *J. Comp. Neurol., 377*, 414-422.

Puelles L, Medina L (2002) Field homology as a way to reconcile genetic and developmental variability with adult homology. *Brain Res. Bull., 57*, 243-255.

Puelles L, Rubenstein JLR (1993) Expression patterns of homeobox and other putative regulatory genes in the embryonic mouse forebrain suggests a neuromeric organization. *TINS, 16*, 472-479.

Puelles L, Rubenstein JLR (2003) Forebrain gene expression domains and the evolving prosomeric model. *TINS, 26*, 469-476.

Puelles L, Amat JA, Martinez-de-la-Torre M (1987) Segment-related, mosaic neurogenetic pattern in the forebrain and mesencephalon of early chick embryos. I. Topography of AChE-positive neuroblasts up to stage HH18. *J. Comp. Neurol., 266*, 147-168.

Puelles L, Kuwana E, Puelles E, Rubenstein JLR (1999) Comparison of the mammalian and avian telencephalon from the perspective of gene expression data. *Eur. J. Morphol., 37*, 139-150.

Puelles L, Kuwana E, Puelles E, Bulfone A, Shimamura K, Keleher J, Smiga S, Rubenstein JLR (2000) Pallial and subpallial derivatives in the embryonic chick and mouse telencephalon, traced by the expression of the genes *Dlx-2*, *Emx-1*, *Nkx-2.1*, *Pax-6*, and *Tbr-1*. *J. Comp. Neurol., 424*, 409-438.

Rakic P (1971) Neuron–glia relationship during granule cell migration in developing cerebellar cortex. A Golgi and electron microscopic study in *Macacus rhesus*. *J. Comp. Neurol., 141*, 283-312.

Rakic P (1972) Mode of cell migration to the superficial layers of fetal monkey neocortex. *J. Comp. Neurol., 145*, 61-84.

Rakic P (1974) Neurons in Rhesus monkey visual cortex: systematic relation between time of origin and eventual disposition. *Science, 183*, 425-427.

Rakic P (1988) Specification of cerebral cortical areas. *Science, 241*, 170-176.

Rakic P (2002) Evolving concepts of cortical radial and areal specification. In: Azmitia E et al. (Eds.), *Progress in Brain Research*, Vol. 136, pp. 265-280. Elsevier, Amsterdam.

Rakic P (2003a) Elusive radial glial cells: historical and evolutionary perspective. *Glia, 43*, 19-32.

Rakic P (2003b) Developmental and evolutionary adaptations of cortical radial glia. *Cereb. Cortex, 13*, 541-549.

Ramon y Cajal S (1890) Sur l'origine et les ramifications des fibres nerveuses de la moelle embryonnaire. *Anat. Anz., 5*, 85-95, see also 111-119.

Reifers F, Böhli H, Walsh EC, Crossley PH, Stainier DYR, Brand M (1998) *Fgf8* is mutated in zebrafish *acerebellar* (*ace*) mutants and is required for maintenance of midbrain-hindbrain boundary development and somitogenesis. *Development, 125*, 2381-2395.

Rendahl H (1924) Embryologische und morphologische studien über das Zwischenhirn beim Huhn. *Acta Zool. (Stockholm), 5*, 241-344.

Rétaux S, Rogard M, Bach I, Failli V, Besson M-J (1999) *Lhx9*: a novel LIM-homeodomain gene expressed in the developing forebrain. *J. Neurosci., 19*, 783-793.

Reyes R, Haendel M, Grant D, Melancon E, Eisen JS (2004) Slow degeneration of zebrafish Rohon-Beard neurons during programmed cell death. *Dev. Dyn., 229*, 30-41.

Rohr KB, Barth KA, Varga ZM, Wilson SW (2001) The Nodal pathway acts upstream of Hedgehog signaling to specify ventral telencephalic identity. *Neuron, 29*, 341-351.

Ross SE, Greenberg ME, Stiles CD (2003) Basic helix–loop–helix factors in cortical development. *Neuron, 39*, 13-25.

Salas C, Broglio C, Rodríguez F (2003) Evolution of forebrain and spatial cognition in vertebrates: conservation across diversity. *Brain Behav. Evol., 62*, 72-82.

Sauer FC (1935) Mitosis in the neural tube. *J. Comp. Neurol., 62*, 37-405.

Schmidt A, Roth G (1993) Patterns of cellular proliferation and migration in the developing tectum mesencephali of the frog, *Rana temporaria* and the salamander *Pleurodeles waltl*. *Cell Tissue Res., 272*, 273-287.

Scholpp S, Brand M (2003) Integrity of the midbrain region is required to maintain the diencephalic–mesencephalic boundary in zebrafish no isthmus/*pax2.1* mutants. *Dev. Dyn., 228*, 313-322.

Schuurmans C, Guillemot F (2002) Molecular mechanisms underlying cell fate specification in the developing telencephalon. *Curr. Opin. Neurobiol., 12*, 26-34.

Shimamura K, Hartigan DJ, Martinez S, Puelles L, Rubenstein JLR (1995) Longitudinal organization of the anterior neural plate and neural tube. *Development, 121*, 3923-3933.

Shimamura K, Martinez S, Puelles L, Rubenstein JLR (1997) Patterns of gene expression subdivide the embryonic forebrain into transverse and longitudinal zones. *Dev. Neurosci.*, *19*, 88-96.

Simeone A, Acampora D, Gulisano M, Stornaiuolo A, Boncinelli E (1992a) Nested expression domains of four homeobox genes in developing rostral brain. *Nature*, *358*, 687-690.

Simeone A, Gulisano M, Acampora D, Stornaiuolo A, Rambaldi M, Boncinelli E (1992b) Two vertebrate homeobox genes related to the *Drosphila empty spiracles* gene are expressed in the embryonic cortex. *EMBO J.*, *11*, 2541-2550.

Smart IH (1976) A pilot study of cell production by the ganglionic eminences of the developing mouse brain. *J. Anat.*, *121*, 71-84.

Smart IH (1985) Differential growth of the cell production systems in the lateral wall of the developing mouse telencephalon. *J. Anat.*, *141*, 219-229.

Smart IHM, Dehay C, Giroud P, Berland M, Kennedy H (2002) Unique morphological features of the proliferative zones and postmitotic compartments of the neural epithelium giving rise to striate and extrastriate cortex in the monkey. *Cereb. Cortex*, *12*, 37-53.

Smith-Fernandez A, Pieau C, Repérant J, Boncinelli E, Wassef M (1998) Expression of the *Emx-1* and *Dlx-1* homeobox genes define three molecularly distinct domains in the telencephalon of the mouse, chick, turtle and frog embryos. *Development*, *125*, 2099-2111.

Sommer L, Ma Q, Anderson DJ (1996) Neurogenins, a novel family of atonal-related bHLH transcription factors, are putative mammalian neuronal determination genes that reveal progenitor cell heterogeneity in the developing CNS and PNS. *Mol. Cell. Neurosci.*, *8*, 221-241.

Stenman J, Toresson H, Campbell K (2003) Identification of two distinct progenitor populations in the lateral ganglionic eminence: implications for striatal and olfactory bulb neurogenesis. *J. Neurosci.*, *23*, 167-174.

Stoykova A, Gruss P (1994) Roles of *Pax*-genes in developing and adult brain as suggested by expression patterns. *J. Neurosci.*, *14*, 1395-1412.

Stoykova A, Treichel D, Hallonet M, Gruss P (2000) *Pax6* modulates the dorsoventral patterning of the mammalian telencephalon. *J. Neurosci.*, *20*, 8042-8050.

Sussel L, Marín O, Kimura S, Rubenstein JLR (1999) Loss of *Nkx2.1* homeobox gene function results in ventral to dorsal molecular respecification within the basal telencephalon: evidence for a transformation of the pallidum into the striatum. *Development*, *126*, 3359-3370.

Tamamaki N, Nakamura K, Okamoto K, Kaneko T (2001) Radial glia is a progenitor of neocortical neurons in the developing cerebral cortex. *Neurosci. Res.*, *41*, 51-60.

Tissir R, Lambert de Rouvroit C, Sire JY, Meyer G, Goffinet AM (2003) Reelin expression during embryonic brain development in *Crocodylus niloticus*. *J. Comp. Neurol.*, *457*, 250-262.

Toresson H, Potter S, Campbell K (2000) Genetic control of dorsal–ventral identity in the telencephalon: opposing roles for *Pax6* and *Gsh2*. *Development*, *127*, 4361-4371.

Torii M-A, Matsuzaki F, Osumi N, Kaibuchi K, Nakamura S, Casarosa S, Guillemot F, Nakafuku M (1999) Transcription factors Mash-1 and Prox-1 delineate early steps in differentiation of neural stem cells in the developing central nervous system. *Development*, *126*, 443-456.

Trevarrow B, Marks DL, Kimmel CB (1990) Organization of hindbrain segments in the zebrafish embryo. *Neuron*, *4*, 669-679.

Vaage S (1969) The segmentation of the primitive neural tube in chick embryos (*Gallus domesticus*). A morphological, histochemical and autoradiographic investigation. *Ergebn. Anat. Entwicklungsgesch.*, *41*, 1-88.

von Baer KE (1828) *Über die Entwickelungsgeschichte der Thiere. Beobachtung und Reflexion*. Bornträger, Königsberg.

von Frowein J, Campbell K, Götz M (2002) Expression of *Ngn1*, *Ngn2*, *Cash1*, *Gsh2* and *Sfrp1* in the developing chick telencephalon. *Mech. Dev.*, *110*, 249-252.

von Kupffer C (1906) Die Morphogenie des ZNS. In: Hertwig O (Ed.), *Handbuch der vergleichenden und experimentellen Entwicklungslehre der Wirbeltiere*, Vol. 2, Part 3, pp. 1-272. Fischer, Jena.

Voogd J, Nieuwenhuys R, van Dongen PAM, ten Donkelaar HJ (1998) Mammals. In: Nieuwenhuys R, ten Donkelaar HJ, Nicholson C (Eds.), *The Central Nervous System of Vertebrates*, Vol. 3, pp. 1637-2097. Springer, New York.

Walther C, Gruss P (1991) *Pax-6*, a murine paired box gene, is expressed in the developing CNS. *Development*, *113*, 1435-1449.

Wang VY, Zoghbi HY (2001) Genetic regulation of cerebellar development. *Nat. Rev. Neurosci.*, *2*, 484-491.

Wang X, Chu LT, He J, Emelyanov A, Korzh V, Gong Z (2001) A novel zebrafish bHLH gene, *neurogenin3*, is expressed in the hypothalamus. *Gene*, *275*, 47-55.

Waseem NH, Lane DP (1990) Monoclonal antibody analysis of the proliferating cell nuclear antigen (PCNA): structural conservation and the detection of a nucleolar form. *J. Cell Sci.*, *96*, 121-129.

Westerfield M (1995) *The Zebrafish Book*. University of Oregon Press, Eugene, OR.

Westin J, Lardelli M (1997) Three novel *Notch* genes in zebrafish: implication for vertebrate *Notch* gene evolution and function. *Dev. Genes Evol.*, *207*, 51-63.

Wichterle H, Turnbull DH, Nery S, Fishell G, Alvarez-Buylla A (2001) In utero fate mapping reveals distinct migratory pathways and fates of neurons born in the mammalian basal forebrain. *Development*, *128*, 3759-3771.

Wilkinson DG, Krumlauf R (1990) Molecular approaches to the segmentation of the hindbrain. *TINS*, *13*, 335-339.

Wilkinson DG, Bhatt S, Chavrier P, Bravo R, Charnay P (1989a) Segment specific expression of a zinc-finger gene in the developing nervous system of the mouse. *Nature*, *337*, 461-465.

Wilkinson DG, Bhatt S, Cook M, Boncinelli E, Krumlauf R (1989b) Segmental expression of *Hox-2* homeobox genes in the developing mouse hindbrain. *Nature*, *341*, 405-409.

Wilson SW, Ross LS, Parrett T, Easter SS Jr. (1990) The development of a simple scaffold of axon tracts in the brain of the embryonic zebrafish, *Brachydanio rerio*. *Development*, *108*, 121-145.

Wullimann MF, Knipp S (2000) Proliferation pattern changes in the zebrafish brain from embryonic through early postembryonic stages. *Anat. Embryol.*, *202*, 385-400.

Wullimann MF, Mueller T (2002) Expression of *Zash-1a* in the postembryonic zebrafish brain allows comparison to mouse *Mash1* domains. *Gene Expr. Patterns*, *1*, 187-192.

Wullimann MF, Mueller T (2004a) Identification and morphogenesis of the eminentia thalami in the zebrafish. *J. Comp. Neurol.*, *471*, 37-48.

Wullimann MF, Mueller T (2004b) Teleostean and mammalian forebrains contrasted: evidence from genes to behavior. *J. Comp. Neurol.*, *475*, 143-162.

Wullimann MF, Puelles L (1999) Postembryonic neural proliferation in the zebrafish forebrain and its relationship to prosomeric domains. *Anat. Embryol.*, *199*, 329-348.

Wullimann MF, Rink E (2001) Detailed immunohistology of *Pax6* protein and tyrosine hydroxylase in the early zebrafish brain suggests role of *Pax6* gene in development of dopaminergic diencephalic neurons. *Dev. Brain Res.*, *131*, 173-191.

Wullimann MF, Rupp B, Reichert H (1996) *Neuroanatomy of the Zebrafish Brain. A Topological Atlas*. Birkhäuser, Basel.

Wurst W, Bally-Cuif L (2001) Neural plate patterning: upstream and downstream of the isthmic organizer. *Nature Rev.*, *2*, 99-108.

Yan YL, Talbo WS, Egan ES, Postlethwait JH (1998) Mutant rescue by BAC clone injection in zebrafish. *Genomics*, *50*, 287-289.

Yun K, Potter S, Rubenstein JLR (2001) *Gsh2* and *Pax6* play complementary roles in dorsoventral patterning of the mammalian telencephalon. *Development*, *128*, 193-205.

Yun K, Fischman S, Johnson J, De Angelis MH, Weinmaster G, Rubenstein JLR (2002) Modulation of the notch signaling by *Mash1* and *Dlx1/2* regulates sequential specification and differentiation of progenitor cell types in the subcortical telencephalon. *Development*, *129*, 5029-5040.

Zeltser LM, Larsen CW, Lumsden A (2001) A new developmental compartment in the forebrain regulated by Lunatic fringe. *Nat. Neurosci.*, *4*, 683-684.

Zerucha T, Stühmer T, Hatch G, Park BK, Long QM, Yu GY, Gambarotta A, Schultz JR, Rubenstein JLR, Ekker M (2000) A highly conserved enhancer in the *Dlx5/Dlx6* intergenic region is the site of cross-regulatory interactions between Dlx genes in the embryonic forebrain. *J. Neurosci.*, *20*, 709-721.

Zhao Y, Marín O, Hermesz E, Powell A, Flames N, Palkovits M, Rubenstein JLR, Westphal H (2003) The LIM-homeobox gene Lhx8 is required for the development of many cholinergic neurons in the mouse forebrain. *PNAS USA*, *100*, 9005-9010.

Index